Statistics Without Tears

A Primer for Non-mathematicians

Derek Rowntree

Statistics Without Tears
A Primer for Non-mathematicians

CHARLES SCRIBNER'S SONS

New York

Copyright © 1981 Derek Rowntree

Library of Congress Cataloging in Publication Data

Rowntree, Derek,
    Statistics without tears.

            Bibliography: p.
            Includes index.
            1. Statistics.   I. Title.
        QA276.12.R68   519.5   82-3157
        ISBN 0-684-17501-0   AACR2
        ISBN 0-684-17502-9

1 3 5 7 9 11 13 15 17 19   F/P   20 18 16 14 12 10 8 6 4 2
1 3 5 7 9 11 13 15 17 19   F/C   20 18 16 14 12 10 8 6 4 2

Printed in the United States of America

Contents

Acknowledgements

This book is dedicated to Professor Peter Moore of the London Business School who gave me just the kind of confidence in approaching statistics that I hope my book will give to its readers.

I take this opportunity also of thanking my Open University colleagues, Professor Brian Lewis, John Bibby, Gordon Burt and Fred Lockwood – and Penguin's advisory editor, David Nelson – for their detailed and constructive comments on earlier drafts which helped me greatly in preparing this final version of the book. Lastly, I thank my secretary, Heida Darling, for producing the immaculate typescript on which those comments were made.

Introduction

So here is yet another introductory book about statistics. Who needs it? What makes me believe I can lead students into the subject along paths that have not already been trodden flat by authors galore?

Perhaps I can best tell you how different this book is going to be by mentioning some of the possible titles I considered for it: *Statistics Without Calculations*, *Statistics for the Innumerate*, *Statistics in Words and Pictures*, *The Underlying Ideas of Statistics*, and *How to Think Statistically*.

All of these say something about my attitudes and intended approach. The first two titles recognize that many non-mathematicians take (are made to take?) a course in statistics as part of some wider program of studies. Such non-mathematical students may be more than a little dismayed at the sight of pages bristling with formulae, equations and calculations. As it happens, many statistics textbooks begin by promising that 'the reader will need no more than the ability to add, subtract, multiply and divide, and to substitute numbers for letters in formulae.' But most soon go on to lose him (or her) in calculations so involved and time-consuming that the student forgets about the concepts that the calculations are supposed to illustrate. He can't see the statistical wood for the computational trees.

The third title expresses *how* I intend to teach the essentials of statistics without computation – by relying on the power of words and pictures. The last two titles indicate where I believe the emphasis should lie in an introductory text – not on calculations but on *ideas*. This book does not teach you how to chew up numbers and spit out plausible 'answers' (a pocket calculator could do this for you), but how to understand the key concepts of statistics and use them in *thinking statistically* about whatever real-world problems you find them relevant to. If you are a 'consumer' of statistics (interpreting other people's research reports, for instance), this may be all you need. If, however, you are a 'producer' of statistics, there are plenty of books to show you how to perform all the necessary calculations. And you should find the learning of such computational procedures much more meaningful after having first acquired a conceptual picture of statistics *as a whole*. For such readers, this book will serve as a useful primer, and that is the idea I have embodied in my sub-title.

But as you can see, I ended up with the main title of *Statistics Without Tears*, from which you may infer that I hope to save you from the weeping (and/or wailing and/or gnashing of teeth) that is so often to be heard among students drawn willy-nilly into the study of statistics. But – be warned – I do *not* imply Statistics Without Effort. As you'll see, I expect you to work for your understanding!

How to use this book

You will soon notice (if you haven't already) that this book does not present you with the usual stream of continuous text. Instead of offering thousands of words for you to 'read, learn and inwardly digest' as best you can, every now and again I pause to ask you a question. This questioning approach is meant to make the book less like a lecture than a 'tutorial in print' – a technique used with considerable success in the correspondence texts produced by Britain's Open University, and in some modern textbooks.

The questions I ask will not, of course, require you to perform calculations. (The nearest you will normally come to that is when I ask you to judge whether something we have discussed is likely to be bigger or smaller than, or about the same size as, something else.) Rather, my questions will require you to recognize examples of the concepts we are discussing, to distinguish between examples and non-examples of a given concept, to relate one concept to another, to apply the concepts to practical problems, to predict likely outcomes, to interpret results, and so on. In short, my questions will be asking you not just to read, but also to *think* about the subject.

So, every now and again, whenever we have got to a point worth emphasizing, you will see a row of asterisks across the page, like the one a few lines below. And just before you get to that row of asterisks you will find I have posed a question of some kind. Here is a question now (a 'multiple-choice' question, this one) to get you into the spirit of the thing.

Judging by what I have just said, what do you think will be my main purpose in asking you questions? Is it:

(a) to slow up your reading? or

(b) to help you learn? or

(c) to test what you have learned?

*   *   *   *   *   *   *

Immediately after the row of asterisks, you will read my comments on possible answers to the question, before we continue with the discussion. So, as far as the question above is concerned, (b) is the answer I hope you'll have chosen. However, it is also true that my questions will slow you up, because they will require you to think as well as to read; in some cases, they will also enable you to test what you have learned so far. However, their main purpose is to ensure that you do learn – by *using* the ideas we discuss.

So the row of asterisks is meant to act as a visual signal reminding you of this message: PLEASE PAUSE HERE. DO NOT READ ON UNTIL YOU HAVE ANSWERED THE QUESTION. Usually, the questions will require only a few seconds' thought, and you will rarely need to write anything down. But do please try to answer each one before going on to read my discussion of it. If you were to skip the questions, you would get through the book a lot faster, but you wouldn't have learned much. In fact, it is quite a good idea to guard against even accidentally seeing my discussion of an answer before trying the question. You may like to keep a sheet of paper handy with which to cover the text below the next row of asterisks, until you are ready to read it.

Four more points about reading this book, before you begin: (1) The subject-matter is largely sequential. That is to say, the ideas in Chapter 2 build on those in Chapter 1; Chapter 3 depends on Chapter 2; Chapter 4 cannot be understood unless you understand the concepts in Chapter 3, and so on. So please read the chapters in the order in which I have set them out, and do not skip any.

(2) Because this sequential style of argument may be rather new to you (especially if your main interest lies on the arts-side or in a social science), you may find it useful, occasionally, to review some earlier chapters before starting on a new one. You should be helped in this by my use of sub-headings for sections within chapters. In each chapter I have also signalled the introduction of important new concepts by printing their names in CAPITAL LETTERS.

(3) Even within a chapter, you will sometimes find it helpful (indeed essential) to read a section or paragraph several times. Don't be alarmed if you do this often. Many of the ideas in this book are very difficult to grasp without repeated reading (and reflection).

(4) For maximum effect, discuss the statistical ideas you meet in this book with tutors and fellow students, or with anyone knowledgeable about the subject. And keep your eyes open for occasions when you can put the ideas to practical use.

1. Statistical inquiry

Let me begin by assuring you that the kind of thinking involved in statistics will not be entirely new to you. Indeed, you will find that many of your day-to-day assumptions and decisions already depend on it.

Take a simple example: suppose I tell you that two adult friends of mine happen to be sitting in the next room while I write. One is five feet tall and the other is just under six feet tall. What would be your best guess as to each one's sex, based on that information alone?

* * * * * * *

I expect you felt fairly confident in assuming that my friend five feet tall was female while the near-six-footer was male. You could have been wrong, of course, but experience tells you that five-foot men and six-foot women are somewhat rare. You have noticed that, by and large, men tend to be taller than women. Of course, you have not seen all men, or all women, and you recognize that many women are taller than many men; nevertheless you feel reasonably confident about generalizing from the particular men and women you have known to men and women as a whole. That is, *in the absence of any other information*, you would think it more likely that a tall adult is male and a small adult is female.

The above is a simple, everyday example of statistical thinking. I could quote many more. Any time you use phrases like: 'On average, I cycle about 100 miles a week' or 'We can expect a lot of rain at this time of year' or 'The earlier you start reviewing, the better you are likely to do in the exam,' you are making a statistical statement, even though you may have performed no calculations. In the first example, past experience is being *summarized* in a rough-and-ready way. In the second and third cases, the speaker is *generalizing* from his previous experience – of how weather varies with the calendar or how exam success is related to study tactics – in order to make a *prediction* for a particular year or a particular student.

Making sense of experience

It is by making sense of our experience that we human beings grow wiser and gain greater control over the environment we live in. This has been true for the development of the human race over the centuries. It is equally true for each of us as individuals in our own lifetimes. Fortunately, we have this capacity for noticing things. We observe people, things and events in the world around us. We notice their similarities and differences, their patterns and regularities – especially when such features could endanger us or, alternatively, be turned to our advantage.

Many of our observations involve us in counting and measuring. Perhaps we do so in rough-and-ready fashion, and often so intuitively that we are scarcely aware of this habit of 'quantification.' Nevertheless our observations and comparisons are often in terms of 'how much?', 'how big?', 'how often?', 'how far?', 'how difficult?', 'how quickly?', 'how well?', and so on.

Sometimes our observations concern a single thing or person or event. For example, we may notice the size of the potato-crop in a particular field this year. We may make several observations about the same thing: not only the size of the crop in this field but also how much fertilizer was used, the nature of the soil, how much sunshine and rain it had, etc. Sometimes our observations concern several similar but different things. For example, we may observe the size of the potato-crop in several different fields this year, or in the same field over a succession of years.

Thus, we may make one or more observations on one individual, or we may do so for several individuals. Soon we have a *collection* of observations (or 'DATA,' to use the technical jargon).

Inquisitively, as if by instinct, we start looking for connections and patterns, similarities and differences, among the things we happen to have noticed. We ask ourselves questions about the data.

For example, what questions might we ask in looking for connections among the data we have collected about the size of potato-crops?

* * * * * * *

We might ask: is the size of the crop similar in all fields this year? Or, is it similar in this field from one year to another? If not, why not? What else is different about those fields, or years, that might explain the differences?

All such questions lead to an even more vital one: what can we learn from the connections we see among this collection of data that might help us act more effectively in the future?

This is where statistics comes in. It has been developed as a way of making sense of collections of observations. It aims, particularly, to help us avoid jumping to conclusions and to be cautious about the extent to which we can *generalize* from our always limited experience.

The tendency to generalize is an essential part of our everyday thinking. Because this particular field was generously treated with a certain fertilizer and gave a bigger than usual potato-crop, we may feel inclined to generalize and suggest that, therefore, *other* fields so treated would give bigger than usual potato-crops.

Would you think it safe to generalize in this way – on the basis of experience with one field? Why, or why not?

* * * * * * *

In fact, such a generalization would be rather dangerous – it is very likely to be wrong. The bigger crop may be due not to the fertilizer but to, say, the weather. (That is, we may have jumped to an incorrect conclusion.) So even the same field, treated in the same way with fertilizer, may give a very different yield in another year. And as for the other fields, they may differ in yet other ways that could influence the potato-yield, e.g. type of soil, crop grown in the previous year, prevalence of plant-disease in neighboring fields, and so on. (Hence the weakness in our generalization.)

So, what is true of one field in one year may not be true of the same field in other years, let alone of other fields. If we want to generalize more confidently, we need more experience – more observa,ions. The more fields we look at, and over more and more years, the more confident we can be in suggesting how the potato-crop is likely to turn out in other, similar fields.

But notice the word 'likely' in the sentence above. 'Likelihood' or 'weighing up the chances' (that is, PROBABILITY) is central to the statistical view of the world. It recognizes no 100% certainties, especially when dealing with individual people, things or events. For example, a particular kind of field may, *in general*, produce a bigger potato-crop if treated in a certain way, but there will be many exceptions.

In which of these two cases would you think me more likely to be proved correct:

(a) if I predict that fields of a certain type will, in general, produce a bigger crop if treated in such-and-such a way? or

(b) if I predict that any such *particular* field you care to pick out will do so?

*   *   *   *   *   *   *

I'd be more likely to be correct in (a) than in (b). While such fields in general (maybe nine out of ten of them) will behave as expected, I can't be sure that any one particular field you happen to choose will be one of those that do.

As you will learn, statistics helps us to look for reliable regularities and associations among things 'in general' and 'in the long run.' At the same time, however, it teaches us proper caution in expecting these to hold true of any particular individuals. The two chief concerns of statistics are with (1) summarizing our experience so that we and other people can understand its essential features, and (2) using the summary to make estimates or predictions about what is likely to be the case in other (perhaps future) situations. In this book we'll be looking at statistical concepts that enable us to summarize and predict more precisely than we normally would in everyday conversation.

What is statistics?

Before we go any further, we'd better take note, in passing, that the word 'statistics' is used in at least four different senses. First of all, it can indicate, very broadly, a whole *subject* or *discipline*, and everything that gets studied or practiced in its name. Secondly, and more specifically, the term may refer to the *methods* used to collect or process or interpret quantitative data. Thirdly, the term may be applied to *collections of data* gathered by those methods. And fourthly, it may refer to certain *specially calculated figures* (e.g. an average) that somehow characterize such a collection of data. Thus, to illustrate the four meanings in turn, a researcher in a firm's *statistics* department may use *statistics* (statistical methods) to gather and interpret *statistics* (data) about the revenue from sales of a new detergent, and may summarize his findings by quoting the *statistics* of 'average sales per thousand of population' in various towns and 'range of sales revenue from town to town.'

The meaning I shall emphasize in this book is the second of those mentioned above: statistics as a set of *methods of inquiry*. It is these methods that enable us to think statistically – a very powerful way to think – about a variety of situations that involve measurements or observations of quantities.

Statistical thinking (of one kind or another) has a long history. From earliest times, kings and governments have been collecting stat(e)istics about the population and resources of their states. The Domesday Book compiled for William the Conqueror is a comparatively recent example. Even the Old Testament mentions rulers, like the Pharaohs of ancient Egypt, who had a keen interest in data about how many people they had available to build pyramids or fight wars, and how much wealth they could conceivably squeeze out of them in taxation. Today, governments are the most prolific generators of statistics (in the sense of collections of data) : on cost of living, unemployment, industrial production, birth-rates, imports and exports, etc. (See the *Statistical Abstracts of the United States*, several hundred pages of data produced by the government, or the equivalent publications in most other industrialized countries.)

Gamblers, too, have made an indispensable contribution to statistical thinking. To them, and their desire to 'figure out the odds' in games of chance, we owe the theory of probability. Such theory only began to develop in the seventeenth century, largely due to the interest aroused in the French mathematician, Blaise Pascal, by (so it is said) the problems posed by a dice-playing friend. The gambling table proved a good test-ground for a theory concerned with prediction, but probability theory soon began to reveal its explanatory and predictive powers in areas like astronomy, heredity and genetics, business, and even warfare. *

Today, few professional activities are untouched by statistical thinking, and most academic disciplines use it to a greater or lesser degree. Its applications in science, especially the 'biological sciences' like genetics, medicine and psychology, are both numerous and well known. But the physical sciences (e.g. meteorology, engineering and physics) also need statistical methods. And even in the humanities, the dating of ancient fragments of textile or pottery has been revolutionized by the essentially statistical technique of radio-carbon dating; while statistical methods have also been used in literary studies to help decide such questions as whether a particular author wrote a certain work, or at what point in his lifetime it was written. Statistics has developed out of an aspect of our everyday thinking to become a ubiquitous tool of systematic research.

But it is time we got down to discussing what it is about statistical thinking that can lend itself to such a variety of pursuits. Statistics arises out of caution in the face of uncertainty. Statistical thinking is a way of recognizing that our observations of the world can never be totally accurate; they are always somewhat uncertain. For instance, a child we record as being four feet

---

* See *How to Take a Chance* by Darrell Huff (Norton, 1964).

in height will not be exactly that – somewhere between 3 feet 11$\frac{1}{2}$ inches and 4 feet $\frac{1}{2}$ inch maybe, but not exactly four feet. And the chance of inaccuracy is even greater if we use our present observations to estimate what observations elsewhere might reveal. Thus, we might want to use our knowledge that four feet is the average height in this child's class to predict the average height in another class.

In such matters there can be no certainty. But statistics enables us to estimate the extent of our errors. Thus, we may express near certainty that the child's height lies within a range of four feet plus or minus half an inch; or we may calculate that the chances are ninety-nine in a hundred that the average height in another class lies within two inches of four feet.

Descriptive and inferential statistics

You will find that statistics textbooks commonly make a distinction between (1) DESCRIPTIVE STATISTICS (methods used to summarize or describe our observations), and (2) INFERENTIAL STATISTICS (using those observations as a basis for making estimates or predictions, i.e. inferences about a situation that has not yet been observed).

Look again at those three 'everyday' statements I mentioned earlier. Which of them appear(s) 'descriptive' and which 'inferential,' in the sense indicated above?
(i)   'On average, I cycle about 100 miles a week';
(ii)   'We can expect a lot of rain at this time of year';
(iii) 'The earlier you start reviewing, the better you are likely to do in the exam.'

* * * * * * *

Statement (i) is descriptive (an attempt to summarize experience), while (ii) and (iii) go beyond what has been observed, in order to make inferences about what is likely to happen in the future.

The distinction between descriptive and inferential statistics depends upon another: the distinction between *samples* and *populations*.

In statistical jargon, 'POPULATION' does not necessarily refer to a body of people. It may refer to people, but it may equally well refer to white mice, to light-bulbs of a particular brand, to substandard dwellings in inner cities, to meteorites, to future examination results in secondary schools, and so on. The point is that population refers to *all* the cases or situations that the 'statistician' wants his inferences or guesses or estimates to apply to. Thus, different statisticians may be making inferences about the learning ability of (all) white mice; predicting how long all light-bulbs of a particular type are likely to burn; estimating the cost of renovating (all) substandard dwellings; predicting the composition of (all) meteorites; guessing the (total) numbers of candidates passing various examinations, and so on.

Perhaps it is also worth pointing out that the researcher will not be interested in every aspect of members of a population. Rather, he is interested in just some – maybe only one – of the many attributes or characteristics that members might have in common. Thus a psychologist may not be concerned to speculate about the tail-length or litter-size of white mice (though these characteristics might interest other researchers); he is interested simply in their learning ability. Neither might the astrophysicist be interested in predicting the geographical distribution or the size of falling meteorites as well as their composition.

However, even if he is interested in only one characteristic of his population, the researcher will be most unlikely to study all members of it. Usually he has to do the best he can with a SAMPLE – a relatively small selection – from within the population. Often he must do this to save time and expense. For the astrophysicist to tour the world inspecting every meteorite that has ever been known to fall would be prohibitively expensive. Again, an industrial researcher who is estimating the burning-life of a brand of light-bulb by 'testing to destruction' cannot test all the population or there will be none left to sell.

In some cases, it may be logically impossible to study all members of the population. The population may be infinite, or simply not yet available for study. Thus, the psychologist who is studying learning ability in white mice will hope his results, and therefore his inferences, will have some application to all white mice – not just the millions that exist at this moment but also the further millions not yet born. He may even hope his results can be generalized to explain *human* learning. Likewise, the astro-physicist may well use his statistics to generalize not just about the meteorites that have already fallen to earth, or even about those that will fall in the future; he may hope to speculate also about the composition of other objects flying around in space.

All such researchers go *beyond* the available information. They generalize from a sample to the population, from the seen to the unseen. (So do we all, though often in a rather careless, uncontrolled way, when using everyday 'common sense.') This idea of generalizing from a sample applies to research in the arts as well as in the sciences. For example, one would not need to have read everything ever written by, say, D. H. Lawrence and Joseph Conrad before one could begin generalizing about how they compared and contrasted as novelists. One could work from a sample of two or three books by each author.

Anyway, *descriptive* statistics is concerned with summarizing or describing a sample. *Inferential* statistics is concerned with generalizing from a sample, to make estimates and inferences about a wider population. Consider a biologist experimenting with the feeding of chicks. He may report (using descriptive statistics) that a particular sample of 60 chicks, fed a particular compound, grow faster than a similar sample fed on some standard diet. So much (the weight gain) he reports as a fact. But he goes beyond the fact. He uses inferential statistics to suggest that *all* similar chicks (the wider population) would grow faster if given similar treatment.

How safe are such generalizations from a part to the whole? Well, that is largely what statistics is about: quantifying the probability of error. We will be looking at the underlying ideas in subsequent chapters. One thing we can say at this stage, however: the reliability of the generalization will depend on how well the sample mirrors the population. Is the sample truly representative of the population?

For example, suppose you read that an educational researcher has established that a group of 12-year-old American boys learned appreciably more French when taught by 'conversation' than did an otherwise similar group who learned from a textbook. The researcher, like all researchers, is not interested simply in the language acquisition of the particular boys in his sample. He wants to generalize from it to suggest how other people (his wider population) might best learn French. For which of the following possible populations would you think it *most reasonable* to suggest that the 'conversational' method might be as effective as with the sample, and for which of them would you think it *least* reasonable to suggest this:

(i)   all 12-year-old American children;
(ii) all 12-year-old American boys;
(iii) all 12-year-old boys;
(iv) all boys; or
(v)   all learners?

* * * * * * *

What I was really asking there was: 'Of which population was the sample likely to be most/least representative?' So, the superiority of the conversational method that was found in the sample would be *most* likely to be true of population (ii), all 12-year-old American boys. But there would still be a big margin of doubt: how typical was the sample of all 12-year-old boys? Did the boys in the sample, for instance, cover the same cross-section of natural ability, interest in languages, previous experience, etc., as would be found in the population as a whole? Clearly, attempts to generalize to populations less and less like the sample would be more and more doomed to error. Girls may not learn in the same way as boys. Children of other nationalities may not learn as American children do. Adult learners may need different approaches from those of children. Hence, what is true of the sample is *least* likely to be true of population (v), all learners.

Over-generalizing is common enough in everyday conversation. But it is sometimes found in 'scientific' discourse also. A researcher may have failed to notice that the members of his sample differed in some important respect from those of the population he was hoping to generalize to. For example, during the Second World War, gunners in R.A.F. bombers returning from raids were asked from which direction they were most frequently attacked by enemy fighters. The majority answer was 'from above and behind.' Why might it have been unwise to assume, supposing you were a gunner, that this would be true of attacks on R.A.F. bombers in general? (Think of the wider population from which that sample of gunners came.)

*   *   *   *   *   *   *

This was not an easy question. However, the risk of a false generalization lay in the fact that the researcher was able to interview only the *survivors* of attacks. It could well be that attacks from below and behind were no less frequent, but did not get represented in the sample because (from the enemy's point of view) they were successful.

Collecting a sample

The problem here has been called the 'paradox of sampling.' A sample is misleading unless it is representative of the population; but how can we tell it is representative unless we already know what we need to know about the population, and therefore have no need of samples! The paradox cannot be completely resolved; some uncertainty must remain. Nevertheless, our statistical methodology enables us to collect samples that are likely to be as representative as possible. This allows us to exercise proper caution and avoid over-generalization.

In some statistical inquiries, the sample 'chooses itself.' (Some writers prefer to use the term 'batch' for such samples.) Suppose, for instance, that a family doctor wants to do some research on the frequency with which various ailments occur among the patients who happen to visit his office over a period of time. Since his sample members choose themselves, he will have to be very careful if he wishes to generalize to any wider population. Certainly he can't assume that the varieties of ailment found among his sample will be typical of people in general living in his area. He might be somewhat safer in inferring that they are similar to those presented to other *local* doctors. He would be much less safe in assuming that doctors in quite different areas would have to deal with the same ailments in the same proportions. In short, he must say to himself: 'I have some interesting data about a sample: but is there a wider population I can generalize to?'

In other inquiries, however, the researcher starts off with a population in mind. He then *selects* a sample that he believes will represent it. In order for it to be representative, members for the sample must be chosen at RANDOM from the population. That is, each member of the population should have an equal chance of being chosen for the sample. This is not always easy to achieve. For example, if you go into a busy street to interview a 'random' sample of passers-by about, say, their political opinions, you are unlikely to be successful. Why? Because you are most likely to approach people who look approachable and are obviously not in a great hurry to be somewhere else. Such people may well differ in their political opinions from people who look surly or bad-tempered, or who are in a hurry. Thus you will have introduced an imbalance or BIAS into your sample. It will not be random.

To avoid bias, it is preferable to use *mechanical* methods of selecting a random sample. For instance, we may give each item in the population a number, and then refer to what is called a table of random numbers. This table will give us as many numbers as we want – a random selection – say 04, 34, 81, 85, 94, 45, 19, 38, 73, 46, if we want a sample of ten. We then pick out those members of the population that have those numbers. This type of sampling can be done effectively through using a computer, especially when large numbers are involved.

If you have neither a computer nor tables of random numbers, all you can do is use the 'raffle' principle. If your population is small enough, put a marker bearing the name or number of each member into a box, shake them up, and then draw out at random (for instance, blindfold) enough markers to make up your sample.

Achieving randomness in our street interviews might be rather more difficult. After all, we don't know who's going to be around, and we can't give them numbers in advance. Can you see any way of making the selection mechanical?

* * * * * * *

The sort of thing one might do is to stand near a corner and decide to accost, say, the fifth person to come round the corner after you have concluded the previous interview. Alternatively, you might decide to interview the first person to round the corner after, say, one minute has elapsed. Anyway, the object is to give every member of the population (people on the street at that time of day) an equal chance of being chosen.

However, it is conceivable that you could use random methods and still end up with a biased sample. That is, the sample could, by accident, still turn out to be unrepresentative of the population you hoped to generalize to. For example, suppose you are trying to determine students' attitudes to the food provided in a certain college. There are 1,000 students in the college (600 men and 400 women). You decide to take a sample of 100 and you select it from the college roll by using random numbers. Is it possible that the sample could consist entirely of men or entirely of women?

* * * * * * *

Yes, of course it's possible that the sample could consist entirely of men or entirely of women. After all, there are 600 men and 400 women, and you're taking only 100 students. Similarly, the sample could consist entirely of students from one year. Such samples would be extremely unlikely, of course. But it is almost certain that the sample would be unbalanced to some extent – that is to say, we could hardly expect the sample of 100 students to contain *exactly* 60 men and 40 women, or to reflect exactly the proportions between first-year, second-year and third-year students. This may or may not matter. There may be no reason to suspect that, say, men and women students will differ any more in their opinions about the food than would blue-eyed students and green-eyed students. But if there is a systematic difference in opinion between students in different categories then the sample is biased. We'll not be able to generalize from its overall opinions to those of the population. Instead, we'll have to generalize from women in the sample to women in the population, and from men to men, and so on.

In such cases, it will be preferable to use what is called a STRATIFIED RANDOM SAMPLE. That is, we have realized in advance that different groups within the population (e.g. different sexes, different age-groups, different income-levels) may differ also in the characteristic we are interested in. So we state *in advance* the number of men and women, adults and children, rich and poor, that we need in our sample. We then choose randomly from *within* those groups (or strata) in the population.

Let's end this chapter with a dramatic example of bias in a sample. The sample was one gathered by doctors at a Canadian hospital some years ago, and consisted of several hundred patients on whom the newly-introduced diphtheria vaccine was tried out. At the same time, a 'CONTROL GROUP' of patients was treated, not with the vaccine but with the methods commonly used up to that period. Over several years of trials, it was found that 16% of the vaccinated patients died, while only 8% of the patients treated in the normal way died. The generalization might seem crystal clear: patients are less likely to survive if you give them the vaccine. And yet that same diphtheria vaccine is used today as a routine preventive measure. Why? The answer lies in a bias built into the samples. Can you imagine what it might be?

* * * * * * *

Well, the fact of the matter is that neither the vaccine sample nor the control group truly represented the population of patients on whom the vaccine is now regularly used. No doubt for the most professional of reasons, the doctors had been administering the vaccine only to patients who were very seriously ill; while the traditional treatment was reserved for patients with milder symptoms. Thus, the two samples were biased, in different directions, and neither truly represented the population of all patients.

The possibility of chance bias within a sample is one we'll have to consider again later in this book, when we think further about generalizing from samples to population. Meanwhile, let's look, in Chapter 2, at the terms in which we might describe what we see in a sample.

2. Describing our sample

What do we do with the members of a sample, once we've got them? As far as statistics is concerned, whatever else we do with them, we are going to produce *a set of numbers* related to whichever of their common characteristics we are interested in. This set of numbers may, for instance, indicate the proportions of married, single, widowed or divorced people among the sample; alternatively, it may indicate their ages or their incomes. If the members of our sample are people, an outraged humanist might suggest that we are *reducing* them to numbers. In fact, it would be bad statistics as well as bad ethics if we allowed ourselves to forget that there is much, much more to the members of such a sample than we have been able to capture in numbers – and it may bear crucially on how we are to interpret our results.

Statistical variables

Samples are made up of individuals. The individuals in a particular sample may be humans, rodents, light-bulbs, months of the year, potato-fields, or whatever. All members of the sample will share some common attribute or characteristic we are interested in: color, sex, weight, price, durability, etc. Furthermore, each individual member of the sample will differ from one or more of the others on this characteristic: some will be one color, some another; some will be male, others female; some will be lighter, others heavier; etc.

So, in looking at the members of a sample, we ask how they *vary* among themselves on one (or more) of such characteristics. Because of the variation among individuals, such characteristics are called variable characteristics or, simply, VARIABLES. A variable in statistics, then, is any attribute or characteristic that will enable us to distinguish between one individual and another.

For instance, suppose you are thinking of buying a second-hand bicycle. Can you list some variables that you might consider in distinguishing between the various bicycles that would be offered for sale?

* * * * * * *

Below, you'll see the variables that I'd be interested in. Your list may include some, or all, of these, and some others:

Make of bicycle (e.g. Raleigh, Falcon, Puch, Fuji, etc.)
Type of bicycle (e.g. racer, tourer, roadster, etc.)
Color
Age
Condition (e.g. Excellent, Acceptable, Poor)
Price
Size of frame
Number of gears

On each of these characteristics, each bicycle could be expected to differ from at least one of the others. They are variable characteristics. ('Number of wheels' would *not* be a variable, because we would not expect to see a bicycle offered for sale with other than two wheels. Variation here is ruled out.)

The question now is: how do we evaluate each individual bicycle, in the sample available to us, in terms of these variables? This depends on the type of variable.

With a variable like 'Make of bicycle,' we set up *categories*, e.g. Raleigh, Falcon, Puch, Fuji, etc. Then we classify or categorize each bicycle simply by noting its name. In this book, I'll be using the term CATEGORY-VARIABLE to refer to any variable that involves putting individuals into categories. In fact, a category-variable like 'Make of bicycle' is often called a 'NOMINAL VARIABLE' (Latin *nominalis* = of a name), because we are giving names to the different forms the variable may take.

Which of the variables in the list above would you say are category-variables?

* * * * * * *

Certainly, 'Type of bicycle' and 'Color' are category-variables. In both cases, we inspect the bicycle concerned and decide which of two or more categories it belongs in. (These are also nominal variables.)

But you may also have considered 'Condition' to be a category-variable. So it is. Three categories of condition are mentioned above: Excellent, Acceptable, Poor. This is a different kind of category-variable, however. The categories here imply that each individual bicycle can be judged as better or worse than others.

Whenever we can say a sample member is better, or bigger, or faster, or in any way has *more* of a certain characteristic than does another, then we can arrange them *in order*. Thus, Excellent–Acceptable–Poor are *ordered* categories, and this kind of category-variable is therefore often called an 'ORDINAL VARIABLE.'

Suppose just ten bicycles are available to us at present. We decide to judge them on 'Relative condition', and we rank them from 1st (the bicycle in best condition) down to 10th (the bicycle in worst condition). Here we have set up ten categories, and we'll have one member of the sample in each. What kind of category-variable is this? Is it (a) nominal or (b) ordinal?

* * * * * * *

'Relative condition' is an *ordinal* variable. We have to arrange the sample members in order, according to how much of the characteristic we judge each to have. Only then can we label them with the rank numbers: 1st, 2nd, 3rd ... 9th, 10th.

But notice here an important difference between splitting the bicycles into three ordered categories and ranking them in order from 1st to 10th. Which method would demand we inspect the bicycles more closely? And which method gives greater information about the individual differences among the bicycles?

* * * * * * *

The method demanding closer inspection, and giving greater information about the differences, is that of *ranking*. With the three ordered categories we might, for example, have lumped five or six bicycles together as 'acceptable.' But with ranking we'd have to find sufficient difference between those five or six bicycles to be able to sort them out into five or six ordered categories (the middle-rank positions).

Before I go on to talk about the other main type of variable I'd like you to be familiar with, let's just notice how *numbers* have been used in ranking. In fact, they were used simply as *labels*: first, second, third, fourth, etc. We can't do arithmetic with these numbers.

Would you, for example, assume that the bicycle ranked 4th was twice as bad as the bicycle ranked 2nd? Would you assume that the difference in condition between bicycles 1 and 2 is equal to that between 3 and 4?

* * * * * * *

Well, I hope you didn't try to make any such assumptions. There are no grounds for them. The numbers are attached simply to show the order – not to show *how much* better each is compared with others. (Really, this is rather like a race in which, for example, the first two runners finish almost together; then there is a gap of a second or two before the third, fourth and fifth cross the line; then the rest of the runners finish with varying intervals of time between them.)

In the second main type of variable I want you to know about, however, numbers are used in quite a different fashion. In fact, with this kind of variable, individual sample members can be described *only* in terms of numbers. This means that when we look at an individual to see how he, she or it differs from others on this characteristic, what we are looking for is a *quantity* that can be counted or measured. In the case of our bicycles, one such variable would be 'Number of gears.'

Here we look at each bicycle and ask: how many? One? Three? Four? Here the numbers are not simply labels. A bicycle with six gears has twice as many as one with three. If I buy one bicycle with five gears and another with ten gears, my two bicycles will have fifteen gears between them. These are numbers (quantities) that you can do arithmetic with.

Look again at the variables listed on page 29. Which ones are of this latest type – where the differences between sample members will be expressed as quantities?

* * * * * * *

The other variables on which bicycles will differ in quantity are: Age, Price, Size of frame. Just as we asked how many gears a bicycle has, so we can ask how old it is; we can ask its price; and we can measure its frame size. All such variables – where what we are looking for is a numerical value, a quantity – we'll call QUANTITY-VARIABLES.

As with category-variables, there is more than one kind of quantity-variable. One distinction we can make is between quantity-variables that do and those that do not use a scale of numbers on which the zero-point means that the characteristic does not exist at all. Most quantity-variables (e.g. income) do have such a zero-point. If someone's income is $0, then he earns nothing; he is completely lacking in income. Similarly, this means that someone who earns $10 per day earns twice as much as someone who earns $5 per day. But some quantity-variables do not have this kind of zero-point – temperature, for instance. An object with a temperature of 0° is *not* completely lacking in heat. Consequently, an object whose temperature is 10° is not twice as hot as one whose temperature is 5°; it is simply 5° hotter. Many quantity-variables invented by social scientists, e.g. intelligence quotient and socio-economic status, are of this latter type.

Another distinction is between quantity-variables that are discrete and those that are continuous. A DISCRETE VARIABLE is one in which the possible values are clearly separated from one another. A classic example is family size: a family can have 1 child, or 2, 3, 4, 5, etc. But it cannot have $2\frac{1}{2}$ or 4.75.

With CONTINUOUS VARIABLES, on the other hand, whatever two values you mention, it is always possible to imagine more possible values in between them. Height is a good example. A child may be 4 feet high this year and 4 feet 3 inches next year. In the meantime, however, he will have been not just 4 feet 1 inch

and 4 feet 2 inches, but an infinite number of heights in between: 4.0001 feet . . . 4.00015 feet . . . 4.0002 feet . . . and so on.

How do we evaluate a sample member in terms of these two
types of quantity-variable? Discrete variables imply *counting*,
but with continuous variables we need a *measurement*. Among
our bicycle variables, 'Price' is discrete. Money is counted rather
than measured (except perhaps in banks where coins are sometimes weighed in bags). But the point is this: you may be asked
for a price, say, of $40 or (for some unlikely reason) of
$40.01; but there's no possible price between these two, or
between any other two that are 1¢ apart. 'Size of frame,' on
the other hand, is continuous: the frame may measure 21
inches or 21.1 inches, or an infinite number of possible sizes
in between.

Which of the other quantity-variables to do with bicycles (page
29) is discrete and which continuous?

* * * * * * *

The only remaining discrete variable is 'Number of gears.' If
we count the gears on a bicycle, we find it has 1, 3, 4, 5, 8 or 10.
In-between values are not possible. (This is rather an unusual
discrete variable, in that its possible values are not equally
spaced.) 'Age,' on the other hand, is a continuous variable: it
is measured along a scale that allows for an infinite number of
possible ages between, say, ten months and eleven months (or
any other two values).

Here is a diagram to illustrate the main relationships between
types of variable:

Different variables produce data that have to be handled differently in statistics, as you'll soon see. The main distinction
you'll need to remember, however, is between what I've called

category-variables and *quantity-variables*. But before we move on, it's worth noticing that data about quantity-variables *can* be converted into category data. For example, people under 5 feet can be categorized as 'small,' those between 5 feet and 6 feet as 'medium,' and those over 6 feet as 'tall.' Similarly, students who have scored less than 60% in an examination may be considered to have 'failed,' while those who got 60% or more have 'passed.' The drawback to this sort of procedure, however, is that LOSS OF INFORMATION occurs. If the numbers in categories only are recorded, we lose data about the actual measured heights and the exact examination scores. Such a sacrifice may sometimes be worth while in the interest of making the data easier to handle. However, the sacrifice should not be made without weighing up the pros and cons.

Two other points of terminology are worth clearing up at this point. First of all, some books use the word 'measurement' in connection with all variables – not just with quantity-variables that are continuous. That is to say, they regard categories as a very crude form of measurement, with ordered categories, ranking, discrete quantity-variables and continuous quantity-variables becoming increasingly less crude. Thus, you may often come across the word 'values' (which one might think would be used only in connection with quantity-variables) being used to refer to the various named categories of a category-variable.

Secondly, you'll find I speak frequently of 'OBSERVATIONS' or 'observed values.' By this, I mean each measurement, or count or classification, that is made for each member of the sample. For instance, if we record the ages of a sample of 100 students, we'll have 100 observations. If we record each one's sex as well, we'll have a total of 200 observations (or 200 'observed values,' if we use the terminology just mentioned.)

The word 'observation' might sometimes be used even though we had actually *not seen* the student's age or sex for ourselves (for example, on a birth certificate). We might not even have seen the student at all, but only his or her responses to a *questionnaire*. In such cases, it would really be preferable to talk of '*recorded* values (or data).'

Error, accuracy and approximations

In collecting data from a sample, which kind would you expect, in general, to be *more accurate*?
(a) data we have collected ourselves by observation?
(b) data collected from questionnaire responses? or
(c) would you expect both to be equally accurate?

* * * * * * *

Data we have collected from our own observations (a) are likely to be more accurate than questionnaire data. With questionnaires, there are many reasons why people may give false information. For instance, they may misunderstand the questions; or they may have forgotten the information asked for (for example, how much gas they used in a previous week), and give a 'guesstimate' instead; or, of course, they may choose to record a deliberate lie (say, about their age or income).

However, as I suggested earlier, we cannot expect perfect accuracy in statistical data. Even when we are doing our own counting or measuring, some ERROR is inevitable. Lapses of concentration or the necessity to work quickly can cause us to put a sample-member in the wrong category, to count an item twice, or to leave one out. Also, with measuring, the limitations of our instruments (e.g. the markings on a ruler) mean that we can never record, for example, the *exact* length, only the length to the nearest inch, or tenth of an inch, etc. So, a room recorded as being 10 feet long (to the nearest foot) may measure anything between $9\frac{1}{2}$ feet and $10\frac{1}{2}$ feet in length. (A slightly smaller room would have been recorded as 9 feet long and a slightly larger one as 11 feet long – to the nearest foot.) The value recorded may be in error by anything up to six inches either way. We could indicate the possible error by stating the room's length as 10 feet $\pm$ 6 inches: that is, 'ten feet *plus or minus* six inches .'

Could the room have been measured in such a way as to produce a recorded value containing (i) a possible error smaller than $\pm$ 6 inches, and/or (ii) no error at all?

* * * * * * *

If we'd used a smaller *unit* of measurement, measuring, say, to the nearest inch, we'd have made the possible error smaller. Thus, if we'd measured the room as being 9 feet 10 inches (to the nearest inch), the true length could not lie outside the range 9 feet 9½ inches to 9 feet 10½ inches. So the maximum possible error (the difference between our recorded value and the true value) would be half an inch. But, no matter how much we reduced the unit of measurement (measuring, say, to the nearest tenth of an inch, or to the nearest millimeter), we'd never reduce the possible error to zero. The same goes for weighing, timing, or any other form of measurement. Experimental scientists use great ingenuity in devising more and more precise instruments with which to do their measuring; yet, however sophisticated the instrument, even they must concede that, between true value and observation, some difference (perhaps infinitesimal) must remain.

Greater accuracy, even if practicable at all, will usually be more expensive – in time, if not in money. And it may not be worth the bother. For example, if you were buying carpet for that room we discussed a few moments ago, you'd no doubt need to know its length and width to the nearest inch; but if you were buying paint to decorate its walls, a measurement to the nearest foot (or even yard) would suffice. A more accurate measurement could not affect the amount of paint you'd decide to buy.

Naturally, the size of the possible errors differs from one subject-area to another. In experimental science, they are likely to be minute. In the social sciences, they are much larger. In business and economics, where so much data is collected by questionnaire, with very little check on the responses, errors can be enormous. National figures concerning, say, unemployment in various towns, or output from various industries, may easily be off by 10 to 15%. (This may be one of the reasons why government predictions of future population-sizes and inflation-rates, for example, can turn out to be so very inaccurate.)

Whatever the subject-area, it is sensible to remember that the observations or recorded values are really just APPROXIMA-TIONS to some true values. If we've counted carefully and enjoyed good co-operation from our questionnaire respondents (or if we've measured carefully with a small enough unit of measurement), our data should contain minimal error. The sample-figures will be *accurate enough* for us to base sensible decisions on them. But if our counting or measuring has been careless or crude, or if we have to rely on data that someone else may (for all we know) have collected carelessly or crudely, then the error may be large and we can't be so confident about interpreting them. Furthermore, despite a common belief to the contrary, statistics cannot make a silk purse out of a sow's ear. Once they've been collected and recorded, well or poorly, no amount of statistical manipulation can improve on the accuracy of our data.

In Chapter 3, we'll go on to consider what *can* be done with our sample data, once we've got it.

3. Summarizing our data

When we have finished collecting our data from a sample, we may have pages covered with figures. (These are our 'observations.') Our first problem is to sort them out and summarize them, so that both we and other people will be able to make sense of them.

Tables and diagrams

I expect you have had enough of those second-hand bicycles in the previous chapter, so let's change the subject.* Suppose a college in a large town has been collecting data from a sample of 50 students, covering many variables relating to student health and social and academic activities.

One of the variables is 'Students' methods of transport to college.' Here the obvious way to arrange our data is in the form of a table, showing the frequency with which each category of transport (bus, car, train, etc.) was mentioned. This will be a FREQUENCY TABLE.

A convenient way to build up such a table (e.g., when checking questionnaire-responses) is to make a *tally mark* for each member of the sample alongside the category into which he (or she) falls.

---

* By the way, I hope you will sympathize with me in the impossibility of finding examples that could be of equal interest to the variety of people (psychologists, economists, geographers, engineers, teachers, biologists, sociologists, etc.) who may be reading this book. If the idea of pulse-rates seems boring and irrelevant to your interests, please feel free to imagine that the numbers refer instead to test-scores, cell-thicknesses, crop-yields, work-days lost through industrial disputes, breaking strengths of materials, or whatever does seem appropriate. The same statistical concepts will apply, whatever the figures refer to.

(As you'll see in the table below, we tally the observations by making a vertical stroke for each new member until we get to the fifth, whose stroke goes across the previous four; then we start building up towards the next 'bundle' of five. The resulting bundles are easy to count up afterwards.)

Students' methods of transport to college

By foot	⫽⫽⫽⫽ ⫽⫽⫽⫽ ‖	12
Bicycle	⫽⫽⫽⫽ ⫽⫽⫽⫽ ⫽⫽⫽⫽	15
Motor-cycle	⫽⫽⫽⫽ ‖	6
Car	⫽⫽⫽⫽	5
Bus	⫽⫽⫽⫽ ‖‖‖	9
Train	‖‖	3

Total = 50 students

The 'tally marks' have been added up at the end of each row to give the actual number of students in each category. When dealing with a sample, as here, however, we are usually more interested in proportions – because we wish to estimate the proportions in the total population. Hence it is normal to convert the figures to *percentages*. This I have done in the table that follows, but I have left the original figures in brackets, so that our readers are in no doubt about the size of the sample. I have also tidied things up by rearranging the categories in order of size.

Students' methods of transport to college

Method of transport	% of students using that method	Actual numbers in sample
Bicycle	30	(15)
Foot	24	(12)
Bus	18	(9)
Motor-cycle	12	(6)
Car	10	(5)
Train	6	(3)
		(TOTAL = 50)

What else might we do to show up the difference in proportions here? The obvious thing is to illustrate the figures with a BLOCK DIAGRAM – with each block proportional in height to the number of students in a category:

Students' methods of transport to college

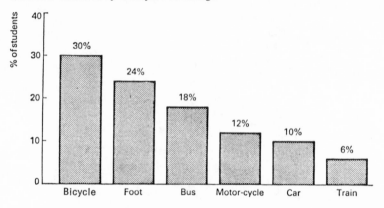

Another way of illustrating such category-data is to use a PIE CHART. Here, a circle is divided into slices, with the angle each makes at the center of the circle being proportional to the frequency in the category concerned.

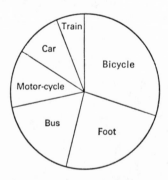

I would say that one of the two types of diagram above is preferable if you want to emphasize how each category compares in size *with the whole*. But the other is preferable if you want to compare the size of one category *with that of each of the others*.

Do you agree? Which type of diagram do you think is preferable for each of these two purposes?

* * * * * * *

I think the pie chart is clearer if we wish to compare each category (slice) with the total. But the block diagram seems clearer if we want to compare one category with another (by comparing the heights of the columns).

Whichever diagram we use, it is apparent that a large proportion of the sample cycles to college. Clearly, if that proportion is true for the 1,000-strong student-population as a whole, then the college will need considerable space for storing bicycles. Again, college staff may wish to compare the figures with those from previous years. For example, the proportion of students travelling by car may be known for each of the last several years up to 1981. When put on a chart, they look like this:

Students travelling to college by car.

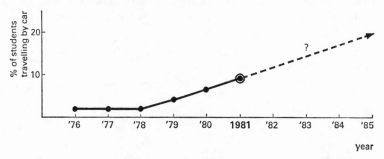

Up to 1978, only about 2% of students travelled by car. Then the proportion began to increase. *If* the trend were to continue after 1981 (as shown by the broken line), about what percentage of students would be travelling to college by car by 1985?

* * * * * * *

If the trend were to continue (and it's a big IF), about 20% of students would be travelling by car by 1985.

Data arranged as in the chart above, recording a succession or series of values of a variable over time, is called a TIME SERIES.

The analysis of time series – looking for trends over a long period of time, and for regular fluctuations within a shorter period – is an important statistical technique in business, economic history, medicine, and a variety of other activities.

Another useful approach to the kind of category-data we have been talking of so far might be to carry out what are called 'cross-tabulations' or 'CROSS-BREAKS.' That is, we break down each category – splitting up the members within it into sub-categories on the basis of what we know about them on some *other* variable we have inquired into, e.g. attendance record, membership of college societies, age, sex, etc. (I expect you have heard about the members of a sample being broken down by age and sex!) Thus, it might emerge that women account for most of the bus-travellers and very few of the cyclists; the car-users are almost exclusively third-year students; and the women who live far away from college are far less likely to be late for early-morning classes than are the men; but the women are less likely to stay on, or make a special journey, to attend college functions; and so on. Such findings may all be of relevance to understanding life within the college.

Now let's see how we might summarize some data from a *quantity-variable*. We have measured the pulse-rate of each of our 50 students. Here are the figures in the order we recorded them:

Pulse-rate (beats per minute) of 50 students

89	68	92	74	76	65	77	83	75	87
85	64	79	77	96	80	70	85	80	80
82	81	86	71	90	87	71	72	62	78
77	90	83	81	73	80	78	81	81	75
82	88	79	79	94	82	66	78	74	72

Presented like this, it's hard to get to grips with such figures. Is there any pattern in them? What is the general picture? How easy is it, for example, to pick out the minimum and maximum pulse-rates? Are the other pulse-rates spread out evenly between the minimum and maximum, or are some pulse-rates much more common than others?

Suppose you know your own pulse-rate to be about 74 beats

per minute; would this seem to be slower or faster than most of these students' pulse-rates? (Don't spend more than half a minute on this.)

* * * * * * *

You no doubt decided that it wasn't at all easy to get that information out of the block of figures. Perhaps you felt it might have been easier if the figures had been arranged in order of size. Let's do that and see whether it helps reveal any pattern:

Pulse-rate (beats per minute) of 50 students

62	64	65	66	68	70	71	71	72	72
73	74	74	75	75	76	77	77	77	78
78	78	79	79	79	80	80	80	80	81
81	81	81	82	82	82	83	83	85	85
86	87	87	88	89	90	90	92	94	96

By the way, when the observed values of a quantity-variable are arranged in order (as they are here, for example) it is called an array or, more generally, a DISTRIBUTION.

Looking at the distribution above, we can now at least see the minimum and maximum pulse-rates: 62 and 96 beats per minute. This gives us the RANGE. In statistics, the range is the difference between the maximum and minimum values you observe in the variable. In this case, 96 minus 62 gives a range of 34 beats per minute.

What else have we gained by arranging the observed values in order of size? One thing we can locate fairly easily is the MEDIAN value. The median (from the Latin for 'middle') is one way we might indicate the center of a distribution. It is often quoted as a kind of representative value.

In fact, the median is whatever value splits a distribution in half. There should be as many observed values greater than the median as there are less. Thus, if the following were the number of miles cycled per week by seven students:

0   16   18   20   33   48   68

the median would be 20 (miles per week). That is the value with equal numbers of observations greater and lesser.

When the number of observed values is even, then of course there will be no observed value with equal numbers of values either side of it. In such cases, the median is quoted as the value half-way between the two middle values. Thus, if the student cycling 68 miles had not been part of the sample and our distribution looked like this:

$$0 \quad 16 \quad 18 \quad 20 \quad 33 \quad 48$$

the median would be half-way between 18 and 20 (the two middle values). $\dfrac{18 + 20}{2} = 19$; and 19 is the value (even though no student actually cycled just 19 miles) that has equal numbers either side of it.

If the two middle observations happen to be of the same value, then that value is given as the median. Thus, if the student who cycled 20 miles had cycled two less, the two middle values would have been 18 and that would have been the median.

What is the median of our distribution of pulse-rates above?

*   *   *   *   *   *   *

Here we have 50 observed values, so we need the value that exceeds 25 of them and falls short of the other 25. We split the difference between the 25th and the 26th value, giving $\dfrac{79 + 80}{2} = 79.5$ beats per minute as the median.

The median is one kind of 'representative' average va'ue (even when it is not a value that was actually observed). But a far more commonly quoted representative value is the ARITHMETIC MEAN. The mean is what is normally called 'the average' in elementary arithmetic (though actually there are several kinds of average besides the arithmetic mean and the median). The mean is calculated by adding together all the observed values and dividing by the number of observations. Thus we would calculate the *mean* number of miles cycled by the seven students mentioned earlier as follows:

$$\frac{0 + 16 + 18 + 20 + 33 + 48 + 68}{7} = \frac{203}{7} = 29 \text{ miles.}$$

Again, no student cycled exactly 29 miles, just as no American family actually has the mean number of children – 2.35!

As I promised you no calculations, I won't set you the daunting task of adding up the 50 pulse-rates and dividing by fifty. I'll simply tell you that the mean pulse-rate is 79.1 beats per minute. Notice that it is not quite the same as the median.

Is your own pulse-rate above or below the 'average' of this sample?

* * * * * * *

Well, I don't know about yours, but mine is well below average for this sample. And this is a very common use of the idea of average. It enables us to locate a particular individual score or value (e.g. one's own pulse-rate) with respect to some value typical or representative of a distribution.

How far have we got in describing our set of data? Arranging the pulse-rates in order of size made it easier to pick out maximum and minimum values and the median value. (It also made it easier for me to calculate the mean, but I won't trouble you with the details.) What has perhaps *not* yet become clear, however, is the overall *shape* of the distribution.

For example, looking back at the distribution, how easy is it to see whether the observed values are spread out fairly evenly over the range or whether they tend to bunch together at any point? (Again, don't spend more than half a minute on this.)

* * * * * * *

Well, it certainly wasn't obvious at a glance. In fact, we can convey a better idea of the pattern of the distribution as a whole by 'picturing' it. The 'dot-diagram' below reveals that the pulse-rates tend to bunch around the middle of the range. For instance,

Each dot represents one student

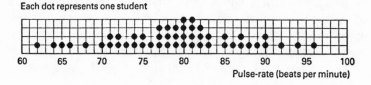

Pulse-rate (beats per minute)

each pulse-rate between 77 and 82 has been recorded for more separate students than has any bigger or smaller pulse-rate.

An arrangement in diagrammatic or tabular form like the one above is called a FREQUENCY DISTRIBUTION It shows up the frequency with which each value in the distribution was observed. For instance, how frequently were pulse-rates of (i) 90 and (ii) 69 beats per minute recorded?

*   *   *   *   *   *   *

Pulse-rates of 90 were recorded on two occasions (frequency = 2) but pulse-rates of 69 were not recorded at all (frequency = 0).

We can use the frequency distribution as pictured in the dot-diagram to check whether any pulse-rate was more frequently recorded than any other. The value in a distribution that has been observed with greatest frequency is called the MODE of the distribution. It is the most 'fashionable' or 'popular' value. Thus, size 8 is the mode, or modal value, in men's shoes – it is the size that shoe shops sell most of. The mode can be another way of indicating a typical or representative value among those recorded. It is a third kind of 'average.'

In the distribution above, no one value has been observed more frequently than any other. There are two values (80 and 81) sharing greatest popularity. We would say the distribution therefore has two modes (80 and 81) and, since they are close together, this might help someone trying to picture it.

In fact, the mode is often most useful as a description when the sample concerns categories rather than quantity-variables. For example, suppose a survey of college teachers showed that more were married than were divorced, single or widowed; then 'married' would be the modal category. (There is no way of calculating a mean or median with category-variables.)

Returning now to our set of pulse-rates, what else can we do to bring out the pattern within the distribution? One thing we can do is to *group* the observations. For example, we can ask how many observations are there of at least 60 but less than 65 . . . at least 65 but less than 70 . . . at least 70 but less than 75 . . . and so on. If we do, we get a table like this:

Pulse-rate (beats per min.)	Number of students (frequency)
60–64	2
65–69	3
70–74	8
75–79	12
80–84	13
85–89	7
90–94	4
95–99	1
	50 = TOTAL

An arrangement like this is called a *grouped* frequency distribution. It brings out the overall pattern even more clearly – in this case, a bunching of observed values in the middle. But it *loses* information about the individual values observed. For instance, the fastest pulse-rate recorded could be 95 or 96 or 97 or 98 or 99 beats per minute. Detail has been sacrificed to clarify the overall pattern.

We can bring out the pattern in a *grouped* frequency distribution even more clearly with a HISTOGRAM. This is a block diagram whose blocks are proportional in area to the frequency in each class or group. In the histogram below, the group '70–74 beats per minute' has twice as many members as the class '90–94' – and so its block is twice as big. (Notice that I have numbered the horizontal scale with the pulse-rate corresponding to the *midpoints* of the classes.)

Pulse-rates of 50 students

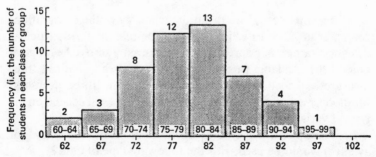

And the *total* area enclosed by the histogram (that is, the shaded area) is proportional to the number of observations (50). If we drew a similar histogram for, say, 25 students on these scales, it would occupy only half the space.

Which of the eight classes in the histogram (60–64, 65–69, etc.) would you say is the *modal* class?

* * * * * * *

The modal class is the one with the largest frequency. In this case it is '80–84 beats per minute' (which has one more member than the class '75–79').

So what have we noticed so far about summarizing the results obtained from a sample? We have seen that a set of 'RAW DATA' – the untreated figures – can be rather obscure. The first step is to rearrange them in order of size. Grouping them may also help emphasize any pattern within the distribution. And diagrams give a better idea of the shape of the distribution than do figures alone.

Also, we have begun to look for figures (like an average, or the range) that quantify important features of the distribution. In fact, to describe a distribution statistically, or to use it in making inferences or predictions, we must have such figures. The two most important of them are a measure of CENTRAL TENDENCY (or average) and a measure of dispersion (or variability).

Central tendency (averages)

By 'central tendency' we mean the tendency of the observations to center around a particular value (or pile up in a particular category) rather than spread themselves evenly across the range or among the available categories. We have already noticed three measures of such a tendency. These are the three averages: the mode, median and mean. Which one is used will depend on the type of variable.

For example, which type of 'measure of central tendency' (or average) would be used with the following type of data?

Method of transport	Number of students
Bicycle	15
Foot	12
Bus	9
Motor-cycle	6
Car	5
Train	3
TOTAL	50

Would it be (a) mode? or (b) median? or (c) mean?

* * * * * * *

The average or measure of central tendency to use with such 'category data' would be the mode. 'Bicycle' would be described as the modal category. There would be no point working out a mean number of students per method of transport (dividing fifty by 6) or in looking for a 'median' method.

While the mode is most likely to be used when our data concerns categories, it will not be much used with quantity-variables. There, the average used will probably be the mean or, occasionally, the median.

As a measure of central tendency, the mean has several advantages, the chief one being that it is fairly stable from one sample to another, that is, if we take a number of samples from the same population, their means are likely to differ less than their medians or their modes. Thus, a sample mean gives us the most reliable estimate of central tendency in the population.

However, there are some situations where it may be as well to concentrate on the sizes of the middle values only and use the median instead of the mean. For example, look at these two distributions of earnings. Each represents a different group of five people:

| X | $30,000 | $38,000 | $42,000 | $57,000 | $73,000 |
| Y | $30,000 | $38,000 | $42,000 | $57,000 | $244,000 |

The median income is the same for each group – $42,000.

But, while the mean income of group X is $48,000, the mean for group Y is $82,200.

Either mean or median could reasonably 'represent' group X. But which average would give the better idea of typical earnings in group Y? Would it be (a) mean or (b) median?

*　*　*　*　*　*　*

In group Y, the median would give a better idea of typical earnings. The mean has been boosted considerably by one very unusual value, while the median remains unaffected.

So the median is preferable in distributions where there are a few extreme (high or low) values observed. The effect of these extreme values would be to distort the mean, pulling it too far away from the center of the distribution.

Also, the median must be used where there is uncertainty about the sizes of some of the values in the distribution. For example, suppose we are given the ages of five bus passengers as follows:

Under 12,　22,　48,　54,　over 65 years.

Here we can't calculate the mean age (unless we arbitrarily assign ages to the youngest and the oldest person). But the median is clearly 48 years. Half the passengers are older and half are younger.

So the mean (or sometimes the median) is the value we normally quote to indicate the center of the distribution. Next, if we are dealing with a quantity-variable, we need a measure of the extent to which the observed values are spread out from this center. The more variable the values, the more dispersed they will be – so we are looking for a measure of DISPERSION (or variability).

Measures of dispersion

To get a picture of dispersion, compare the two dot-diagrams below. The first of them is the one you have already seen.

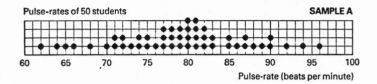

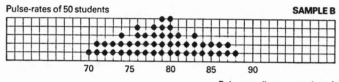

What seems to you the most striking difference between the two distributions? Can you mention a figure (a statistic that you have already met) that would help quantify the difference?

*   *   *   *   *   *   *

To me, the most striking difference is that the observed values in B are far less spread out than they are in A. Perhaps the simplest way of quantifying dispersion in the two distributions is to compare the *ranges*:

$$\text{In A, range} = 96 - 62 = 34 \text{ beats}$$
$$\text{In B, range} = 88 - 70 = 18 \text{ beats}$$

Clearly there is much less variability in sample B than in A.

Now the range is a rough-and-ready measure of dispersion. Its great virtue is that it is so easy to calculate and is fairly apparent at a casual glance. However, the range is not usually a measure we can put much trust in. It depends totally on just two values – the two most extreme cases. These may be quite untypical of any other values in the sample. You've seen this already in the income distribution of group Y on page 49.

We can make the point also with dot-diagrams. Here are the marks gained by two different groups of twenty students on a test:

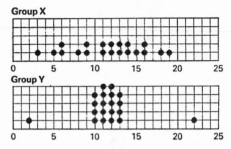

Which of the two distributions above would you tend to say is the more dispersed – group X or group Y? Does it also have the bigger range?

* * * * * * *

Overall, surely, the distribution of group X seems to be the more dispersed. In group Y, apart from the two extreme values, only four different values (all adjacent) – 10, 11, 12, 13 – were observed, whereas thirteen different values were observed in X. However, thanks to the influence of the two extreme values in Y, its range is bigger than that of X.

One way of getting a 'fairer' measure of dispersion is to take a kind of 'mini-range' from nearer the center of a distribution, thus avoiding any extremes. This range is based on what are called the QUARTILES of the distribution. Just as the median is the value that cuts the observations in two, so the quartiles are the values that cut the observations into four equal lots:

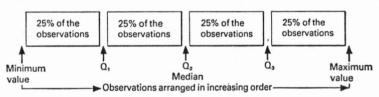

As the diagram shows, there are three quartiles, Q_1, Q_2, and Q_3. The second quartile is the same value as the median.

The 'mini-range' I mentioned above is called the INTER-QUARTILE RANGE. It is the distance between Q_1 and Q_3.

Let's apply this to our two distributions, group X and group Y. Since there are twenty observations in each, we want the value that cuts off the bottom five values and the top five. Thus, Q_1 will be a value half-way between the fifth and sixth observations. Q_3 will be the value half-way between the fifteenth and sixteenth.

In distribution X, the fifth value is 8 and the sixth value is 9. So $Q_1 = 8\frac{1}{2}$. Also, since the fifteenth value is 14 and the sixteenth is 15, $Q_3 = 14\frac{1}{2}$.

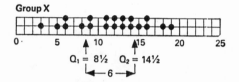

Therefore the inter-quartile range $= 14\frac{1}{2} - 8\frac{1}{2} = 6$ marks.

What is the inter-quartile range for distribution Y?

*   *   *   *   *   *   *

In distribution Y, the fifth value is 10 and the sixth is 11. So $Q_1 = 10\frac{1}{2}$. The fifteenth value is 13 and the sixteenth is 14, so $Q_3 = 13\frac{1}{2}$, and the inter-quartile range $= 13\frac{1}{2} - 10\frac{1}{2} = 3$ marks.

You will no doubt agree that the inter-quartile range gives a more reasonable indication of the dispersion in these two distributions than does the full range.

The inter-quartile range is the measure of dispersion you will often find used along with the median. But it's time we began to concentrate now on the measure of dispersion that is more commonly used than any other: the STANDARD DEVIATION. Like the mean, the standard deviation takes *all* the observed values into account.

How does the standard deviation work? If there were no dispersion at all in a distribution, all the observed values would be the same. The mean would also be the same as this repeated value. No observed value would deviate or differ from the mean. But, with dispersion, the observed values do deviate from the mean, some by a lot, some by only a little. Quoting the standard deviation of a distribution is a way of indicating a kind of 'average' amount by which all the values deviate from the mean. The greater the dispersion, the bigger the deviations and the bigger the standard ('average') deviation.

Which of these sets of values would you expect to have the *larger* standard deviation?

(a)	6	24	37	49	64	(mean = 36)
(b)	111	114	117	118	120	(mean = 116)

* * * * * * *

The values in (a) are more dispersed (that is, they deviate more from the mean) than those in (b), so we can expect the standard deviation to be larger. Let's see how this works out. In (b) the values differ from the mean of 116 as follows:

Value:	111	114	117	118	120
Deviation from 116:	−5	−2	+1	+2	+4

Now we can't simply take an average (arithmetic mean) of the deviations because we'll find they always add up to zero – the negative deviations will exactly cancel out the positive. So, to overcome this difficulty, we *square* each deviation, thus getting rid of the minus signs:

Deviation:	−5	−2	+1	+2	+4
Squared deviation:	25	4	1	4	16

The mean of these squared deviations is called the VARIANCE:

$$\text{Variance} = \frac{25 + 4 + 1 + 4 + 16}{5} = \frac{50}{5} = 10$$

The variance is a measure with uses of its own (as we'll see later in this book). But it does have one disadvantage for everyday practical purposes: if the original values (and therefore the mean) of the distribution were in units of, let's say, 'heart-beats per minute', then the variance would be so many '*square* heart-beats per minute'! This would hardly be an easy idea to play around with. So, to get the measure of dispersion back into the same units as the observed values (and the measure of central tendency), we take the square root of the variance – and this is what we call the standard deviation:

$$\text{Standard deviation of distribution (b)} = \sqrt{10} = 3.16$$

The same calculation for distribution (a) above would give this result:

Value:	6	24	37	49	64	(mean = 36)
Deviation from 36:	−30	−12	+1	+13	+28	
Squared deviation:	900	144	1	169	784	

$$\text{Variance} = \frac{900 + 144 + 1 + 169 + 784}{5} = \frac{1,998}{5} = 399.6$$

$$\text{Standard deviation (a)} = \sqrt{399.6} = 20$$

As I am sure you predicted, we see that the standard deviation of distribution (a) is very much greater than that of (b). This is because distribution (a) is much more dispersed.

Look back now at the two distributions pictured on page 52. Which of those would you expect to have the larger standard deviation?

* * * * * * *

Of the two distributions on page 52, X has a larger standard deviation than Y: 4.3 marks, as compared with 3.3 marks.

Now look back at the two distributions (A and B) pictured on page 51. One of the pairs of figures in the list below shows the standard deviations for the two distributions. Which would you think is the correct pair? Which of the two standard deviations belongs to which of the distributions?

(a) 4.6 and 7.6 beats per minute;
(b) 7.6 and 37 beats per minute; or
(c) 19 and 37 beats per minute.

* * * * * * *

The standard deviation of A is 7.6 and that of B is 4.6 beats per minute. I don't suppose you had any trouble deciding that the larger standard deviation in the pair went with the more dispersed distribution. But was it difficult deciding which was the correct pair? I hope you were able to reject pair (b) by noticing that even if 7.6 were correct for the less dispersed distribution, 37 would exceed even the range in the other. Similarly, the figures given in pair (c) would exceed the range for both distributions.

In fact, the standard deviation never approaches anywhere near the range. Even with a very highly dispersed set of values like:

$$1 \quad 2 \quad 3 \quad 997 \quad 998 \quad 1000$$

where the range is 999, the standard deviation is only about 500. With samples of about ten, the standard deviation can be expected to be about one third of the range. With samples of one hundred, it will be down to about one fifth. These rough ratios may be helpful if you ever need to *guess* a standard deviation.

So how do we summarize the data provided by sample members? Rather than presenting it 'raw' we can:

(1) display it in some kind of table that shows up any pattern within it; and

(2) illustrate it with a diagram that gives a picture of the quantities involved; and

(3) find a suitable figure to indicate its 'central tendency' (e.g. mode, median or mean) and, in the case of a quantity-variable, another figure to indiciate its 'dispersion' (e.g. range, inter-quartile range, or standard deviation).

The type of table, diagram and figure we'd use would depend chiefly on whether the data related to category-variables or to quantity-variables. In the next chapter we'll build on these ideas.

4. The shape of a distribution

With the arithmetic mean and the standard deviation you have two very powerful ways of describing most statistical distributions that involve quantity-variables. (And the median and inter-quartile range can be useful statistics on occasions too.) However, we must not overlook the descriptive power of pictures, particularly in conveying the overall *shape* of the distribution.

It may or may not have struck you that all the distributions we've looked at so far have been fairly *symmetrical*. Indeed, they have all been symmetrical in the same way: most observations being recorded near the center of the range of values and gradually thinning out towards the extremes. Here is one example:

Distribution of pulse-rates among 50 students

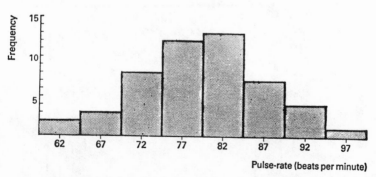

Such symmetry is very common in statistical distributions – especially where biological variations are concerned. But it is not universal.

Skewed distributions

Look at the table below. It contains two grouped distributions (X and Y). Each shows the scores made by the same students on two mathematics tests.

Marks	Number of students	
	Test X	Test Y
0–4	4	0
5–9	10	1
10–14	17	2
15–19	13	5
20–24	11	7
25–29	7	13
30–34	4	19
35–39	2	14
40–44	1	8

Which of the two histograms below, (a) or (b), illustrates each of the distributions X and Y? (I have marked only the group mid-points along the horizontal scales.) How would you describe the difference between the two distributions?

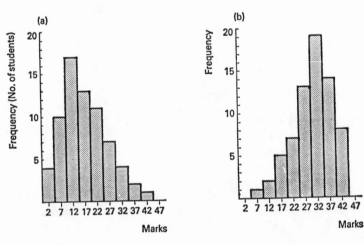

* * * * * * *

The histogram on the left (a) is for scores on test X, while histogram (b) represents the score distribution on test Y.

Clearly, the two distributions are not at all symmetrical Unlike those we've looked at so far, the bulk of the observations are *not* found in the middle of the range of marks (around 20–30) with numbers of observations falling off fairly evenly to either side as the more extreme values are recorded. Rather, in each distribution the bulk of the observations are piled up towards one end of the range and tail off gradually towards the other. But the distributions differ in that each piles up towards a different end.

Such distributions as this are said to be *skewed*. The SKEW is the long tail of observations. (I found it helpful to remember that '*queue*' is French for 'tail' and think of the 'skew' as 's(queue)'.) A distribution can be positively or negatively skewed: positively, if the skew (tail) is on the right (positive) side; negatively, if the skew (tail) is on the left (negative) side.

Which of the two distributions above is positively, and which negatively, skewed?

* * * * * * *

Distribution (a) is positively skewed while (b) is negatively skewed. In (a) the long tail is on the right: in (b) it is on the left.

Although the maximum and minimum observed values in each distribution are much the same, the means are greatly different. In (a) the mean is 17.1 marks; in (b) it is 30.2 marks.*

It is worth noticing the effect that skewness has on the relative sizes and positions of mean, median and mode. In symmetrical distributions of the kind we were looking at earlier, all three measures of central tendency are to be found in the same place – at the center. Look at this simple dot-diagram, for instance:

*In case you're wondering how I calculated the mean of a *grouped* distribution, I operated as though all the observations for a particular group of values were recorded for the value at the *mid-point* of that group. Thus, for example, I treated the eleven observations of '20–24 marks' in (a) as though they were all recorded for a mark of 22. I would operate likewise in calculating the standard deviation. The result will be very little different from what would have been obtained, using the 'raw' marks.

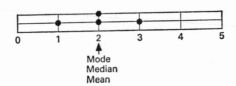

It shows one observation of value 1, two of 2 and one of 3. The mode – the most frequently observed value – is 2. The median – the value that splits the distribution into equal numbers of observations larger and smaller – is 2. And the mean – the total value of the observations shared out equally between them, i.e. $\dfrac{1 + 2 + 2 + 3}{4}$ ... is also 2.

Now let's see what happens if we skew that simple distribution by adding a couple of observations to one side, thus giving it a tail. We will add one observation of 4 and one of 5:

How does this influence the three averages? Well, the most frequently observed value is still 2; so 2 remains the mode. But what about the median? Since there are now six observations (rather than four) in the distribution, the median has to lie midway between the values of the third and the fourth. The third observation is a 2 and the fourth is a 3; so the median is $2\frac{1}{2}$. The median, then, has been dragged away from the mode, in the direction of the skew.

And what about the mean? Where will it lie in relation to the other two averages, now that the distribution is skewed?

* * * * * * *

The mean gets pulled even further in the direction of the skew, because it is influenced by the size of the new observations as well as their number. In fact, the mean of the new distribution is

$$\frac{1 + 2 + 2 + 3 + 4 + 5}{6} = \frac{17}{6} = 2.8$$

So, in the skewed distribution, the position of the three averages is like this:

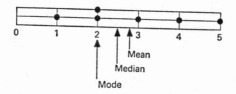

And if we'd had a distribution with a tail on the other side, the positions would have been exactly the reverse:

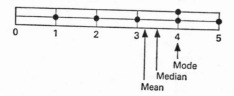

In a skewed distribution, the relative positions of the three averages is always predictable. The mode will be under the peak of the distribution, the mean will have been pulled out in the direction of the skew (whether to left or right) and the median will be in between the mode and the mean. The greater the skewness, the greater will be the distance between the mode and the mean. (There is a statistic known as the 'coefficient of skewness' which is calculated from this relationship, but you probably won't often see it used.)

I'll now show you our two mark-distributions again, this time with arrows indicating where the three averages are in each:

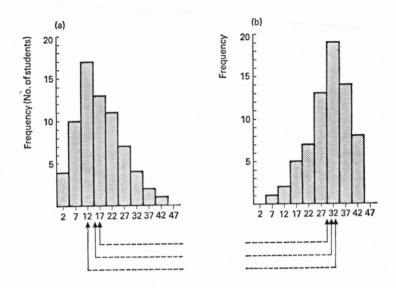

Test your grasp of skewness by pencilling in the names of the three averages on each set of broken lines above.

* * * * * * *

From left to right: in distribution (a) the averages are mode, median and mean; in (b) they are mean, median and mode.

By the way, there is an explanation for the two different kinds of skewness on these two tests. Distribution (a) shows students' marks on a mathematics test soon after lessons began, while (b) shows how they performed at the end of the course. Clearly, the great majority did very much better, although a few students seem not to have benefited and are still in the low-score groups.

Looking back at the histograms, can you form an impression as to whether *dispersion* in distribution (b) has increased, decreased or stayed about the same as in distribution (a)?

* * * * * * *

Well, there really is little obvious difference. However, dispersion in distribution (b) is rather smaller than in distribution (a). Notice that the range is slightly less and the three highest columns slightly higher. In fact the standard deviation in distribution (b) is 8.1 marks, compared with 9.1 marks in (a). (So these summary figures reveal what is not too apparent in the diagrams.)

All distributions you meet in practice are likely to have a certain amount of skewness. It can be even more exaggerated than we have seen in the above example.

For instance, suppose you stood at a busy cross-roads each day for a month and counted the number of traffic accidents each day. You might well find that you recorded 0 accidents on more days than you recorded any other value – followed by 1, 2, 3, 4 . . . accidents, in decreasing order of frequency.

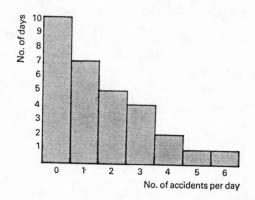

Other everyday examples of highly skewed distributions might be the incomes of salary-earners or the number of faults in each of a batch of newly produced television tubes.

In this book, however, we will be concentrating on the very common kind of distribution that is fairly symmetrical around a central peak. Before we go on to consider this further, though, we ought to notice that there could be other kinds of symmetrical distributions. For instance we could have *bi*modal symmetry as in the histogram below, with two peaks, one towards each end of the range. This might represent the results obtained in an examination by two very different grades of students – really two overlapping distributions.

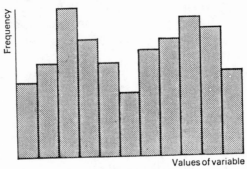

Introducing the normal distribution

Before we talk any further about distributions, let's simplify the way we picture them. Instead of drawing dots or bars, we'll use the outline that defines the boundary of the distribution. If we join up the uppermost dots in a dot-diagram, or the tops of the bars in a histogram, with a smooth curved line, we get what is called the CURVE OF THE DISTRIBUTION. This is the underlying shape of the distribution. For instance, here are our two distributions with their underlying curves:

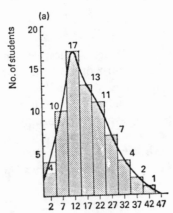

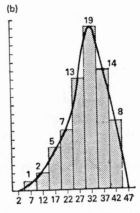

Now look again at the two distributions I showed you on pages 63 and 64, the 'accidents' distribution and the bimodal distribution. Draw a rough sketch of the 'curve' of each distribution.

* * * * * * *

Your curves should look roughly like this:

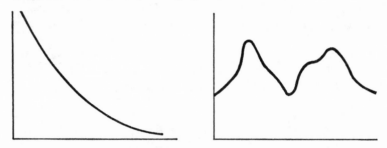

Now let's try the same thing with a dot-diagram. Here again is our distribution of 50 student pulse-rates. As you see, if we try to outline it with a curve, the result is pretty lumpy:

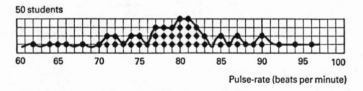

Pulse-rate (beats per minute)

Although the students tend to pile up towards the middle of the range, they don't do so very evenly, and the curve contains a number of 'peaks' and 'valleys.' It is not as smooth as those we have looked at so far.

However, suppose we measured the pulse-rates of 25 more students and added them to our dot-diagram. It would then show the distribution of 75 pulse-rates. We could go on to measure more and more students, so as to show the distribution of, say, 100, 150 or 200 pulse-rates.

As we added to our distribution, we would find the 'peak-and-valley' effect beginning to disappear. As the sample-size increased, the outline curve of its distribution would become smoother and smoother. For instance:

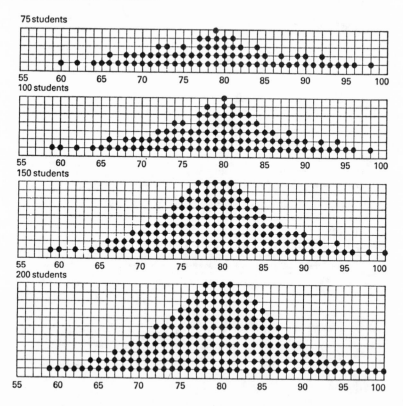

Sketch the outline curve of the distribution shown in that final chart – for 200 student pulse-rates.

*   *   *   *   *   *   *

Your curve should look rather like this:

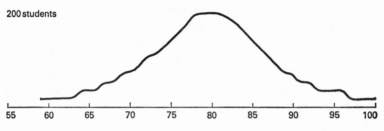

200 students

55   60   65   70   75   80   85   90   95   100

pulse-rate (beats per minute)

Clearly, the outline of this distribution is beginning to look very smooth as we increase the number of students. Indeed, if we measured *thousands* of students, rather than just hundreds, we'd expect to end up with a curve something like this:

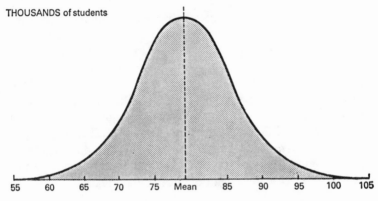

THOUSANDS of students

55   60   65   70   75   Mean   85   90   95   100   105

Pulse-rate (beats per minute)

Imagine, if you like, that the 'dots' in the chart above are so tightly packed that they have merged into one mass. They are no longer distinguishable from one another. (In fact, I have much reduced the vertical scale from that in the previous dot-diagrams. After all, the area under a curve is, like that in a histogram, proportional to the number of observations. So, if I'd kept to the same scale as before, I'd have needed a much taller diagram to represent thousands of students rather than just a couple of hundred.)

The shape of the distribution above follows what is called the NORMAL CURVE of distribution. (This curve was first recognized in the seventeenth century by the English mathematician, de Moivre.) The curve is bell-shaped and perfectly symmetrical; and its mean (and mode and median) is in the center.

As with any distribution curve (normal or not), a line drawn vertically upwards from a value on the base-line will cut the area enclosed by the curve into two slices. With a 'normal' distribution, as you'll soon see, we'll be able to find out what percentage of the observations lie on *either side* of that particular value of the variable.

For instance, in the diagram above, what percentage of the total observations lies on either side of the mean (marked by a vertical broken line)?

* * * * * * *

The mean here is in the center of a perfectly symmetrical distribution. So the broken line cuts the total area into two *equal* slices, so that 50% of the observations lie on one side of the mean and 50% on the other side.

The normal curve may be tall and thin, or short and stocky, or slumping out very flatly. This depends on whether the standard deviation is small or large – or on the relationship between the particular vertical and horizontal scales on which we have chosen to draw the graph. For instance, in the diagram below, notice how the curve gets flatter and flatter as we compress the vertical scale compared with the horizontal scale.

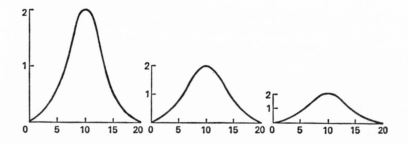

But whatever the height or width of the normal curve, the distribution of the *area* under it is always the same. In the diagram above, the proportion of the total area that lies between, say, 15 and 20 on the horizontal scale is the same under each of the three curves.

Below, I've drawn three different curves (*a*, *b* and *c*), all on the same chart. Which of the three curves looks most like that of a 'normal' distribution?

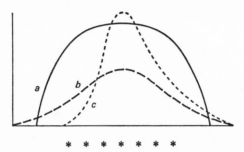

* * * * * * *

Curve *b* looks most like that of a normal distribution. It is not only symmetrical about a central peak (which *c* is not) – but it also appears to have about the right proportions (which *a* does not).

When we call this the 'normal' curve, we do *not* mean that it is the 'usual' curve. Rather, 'norm' is being used in the sense of a pattern or standard – ultimate, idealized, 'perfect' – against which we can compare the distributions we actually find in the real world. And it is a fact that many such distributions (e.g. the sizes or weights of natural or man-made objects of a given type) do come close to this idealized curve. That is to say, they are reasonably symmetrical and bell-shaped about a central mean.

We must remember, though, that the normal curve is a mathematical abstraction – defined by what I'm sure many of my readers would regard as a particularly horrendous-looking equation! – and it assumes an infinitely large population. Samples of the size we get in everyday life will never produce so perfect a curve.

However, quite a small sample can often produce a fairly bell-shaped distribution. This may not have been too apparent in our *dot-diagram* of 50 student pulse-rates (though it was certainly

becoming clear with a sample of 100). But, even with 50, there is
the suggestion of a central peak, with observations gradually (if
unevenly) thinning out to either side:

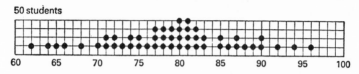

But what if we'd drawn a curve based on the *grouped* distribution? Look back at the *histogram* of pulse-rates (page 57).
Sketch in a curve by connecting the mid-points of the tops of the
blocks. Does this curve look any more 'normal'?

*   *   *   *   *   *   *

As you'll see, the curve based on the *grouped* data is still rather
lumpy and somewhat skewed; but it does bear more of a resem-
blance to the normal curve:

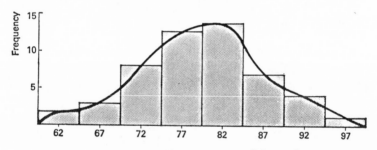

So, even with quite a small real-life sample, the distribution
can look as if it is 'trying' to be normal! This suggests that such
a sample may have come from a larger population whose distribution could indeed be described by the normal curve.

When this is the case, we can interpret the sample, using
certain powerful characteristics of the normal distribution. In
fact, the normal curve is characterized by the relationship

between its mean and its standard deviation. The shape of the normal distribution is such that *we can state the proportion of the population that will lie between any two values of the variable.* To accomplish this, all we need to know is the mean and the standard deviation. We can then regard any given value in the distribution as being 'so many standard deviations' away from the mean. We use the standard deviation as a *unit of measurement.*

For instance, suppose we have a sample of student pulse-rates with a mean of 80 and a standard deviation of 6 beats per minute. A student whose pulse-rate is 86 beats per minute would be 'one standard deviation above the mean.' (Let's abbreviate standard deviation to SD when we're using it as a unit of measurement.) Similarly, a student whose pulse-rate is 74 beats per minute could be said to be 1 SD below the mean. A pulse-rate of 89 beats per minute would be $1\frac{1}{2}$ SD above the mean, and so on.

In terms of mean = 80 and SD = 6, how would you describe the pulse-rates of students who recorded (i) 98 beats per minute, (ii) 77 beats per minute and (iii) 68 beats per minute? How far (in SDs) is each one above or below the mean?

* * * * * * *

With a mean of 80 and a standard deviation (SD) of 6 beats per minute: 98 beats is 3 SD (i.e. 3 × 6) above the mean; 77 beats is $\frac{1}{2}$ SD (i.e. $\frac{3}{6}$) below the mean; and 68 beats is 2 SD (i.e. 2 × 6) below the mean.

If we draw just the base-line of the histogram or curve of the distribution, we can show the positions of the values we've mentioned like this:

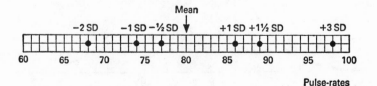

Pulse-rates

Thus *any* value in a distribution can be re-expressed as so many SDs above or below the mean. This we can do whether the distribution is normal or not. But when it is normal (or approximately so), we can use what we know of the normal curve to estimate how many observations lie between any two given values. The standard deviation slices up a bell-shaped distribution into 'standard'-sized slices, each slice containing a *known* percentage of the total observations. You've already noticed that half of them lie on either side of the mean. But, as you'll see below, about two-thirds of them are to be found within one standard deviation either side of the mean – one-third on one side, and one-third on the other.

Proportions under the normal curve

Look at the curve of the normal distribution below as it falls away from its peak on either side. At first the slope is convex (bulging outwards) as it falls faster and faster. But at a certain point – known as the 'point of inflection' (bending back) – it becomes concave (bulging inwards) as the slope begins to level off. The distance between the mean and the point of inflection on either side is equal to the standard deviation. About ⅔ (68%) of all the observations in a normal distribution lie within one standard deviation either side of the mean. We call this range 'Mean plus or minus one SD', or M ± 1 SD.

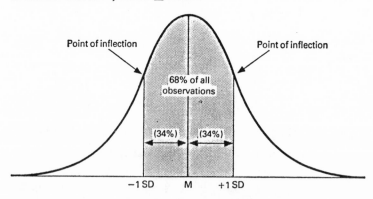

About what percentage of the observations in a normal distribution will have values greater than 1 SD above the mean?

* * * * * * *

We know that M ± 1 SD (the range between 1 SD below the mean and 1 SD above it) encloses about 68% of the observations. So 100 − 68 = 32% must lie outside this range. Since the distribution is symmetrical, we can expect half of the 32% (that is 16%) to lie below −1 SD and the other 16% to lie above +1 SD.

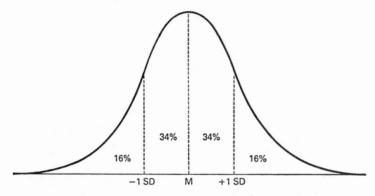

Let me give you an example of how you might use this knowledge. A police radar check has monitored the speeds of 1,000 cars passing a certain point, and the distribution of speeds is approximately normal. Suppose also that the mean speed is 45 mph and the standard deviation 5 mph.

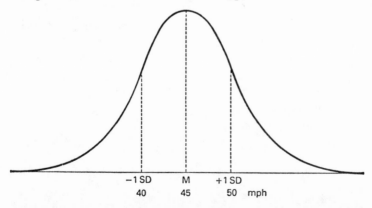

If I tell you that the legal speed-limit at the check-point is 40 mph, can you tell approximately how many cars were breaking the law?

*   *   *   *   *   *   *

The legal speed limit of 40 mph is 1 standard deviation (5 mph) below the mean speed of 45 mph. So the 50% of cars travelling faster than the mean speed plus the 34% between the mean and −1 SD were all breaking the law. 50% + 34% = 84%, and 84% of 1,000 is 840 cars.

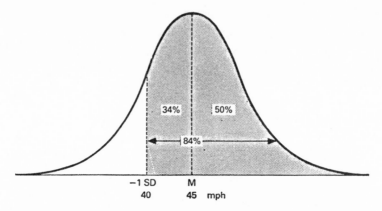

(If the 1,000 cars were a *random* sample of all cars passing the check-point, then we would be justified in making similar estimates about speeding within that population. But more of this in the next chapter.)

So we've seen that the range of values within one standard deviation either side of the mean takes in about two-thirds of the observations. What happens if we extend the range by an amount equal to another standard deviation on either side? We find that 95% of the observations lie within about two (actually 1.96) standard deviations either side of the mean:

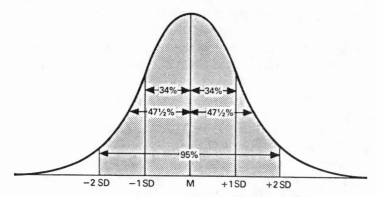

The shaded area represents 95% of the area under the curve. It is bounded on the left by a vertical line drawn up from whatever value of the variable falls at the −2 SD position. On the right it is bounded by the +2 SD point.

Try to see the answers to the following questions in the diagram above:

  (i) About what percentage of the observations will be at least one but less than two standard deviations *below* the mean?

 (ii) What percentage of the observations will be more than two standard deviations *away from* the mean?

(iii) Mark −3 SD and +3 SD on the diagram; about what percentage of the observations would you say lie within three standard deviations of the mean?

* * * * * * *

  (i) About 13½% of the observations will be at least one but less than two standard deviations *below* the mean; i.e. 47½ − 34%.

 (ii) Only 5% of the observations will be more than two standard deviations *away from* the mean (2½% either side).

(iii) Practically 100% of the observations (actually 99.7%) lie within three standard deviations of the mean. (Less than 1% of the observations lie beyond mean ±3 SD. Hence, in the normal distribution, the standard deviation is about equal to one-sixth of the total range.)

In fact, there are tables that will tell us what percentage of the observations are cut off by any given value in a distribution. All we have to do is translate 69 heart-beats per minute, or 51 mph (or whatever) into so many SDs above or below the mean. Then (even with 'awkward' fractions like −1.85 SD or +2.33 SD) we can use the tables to read off the percentage of observations (e.g. of students or of cars) we may expect on either side of that value.

Some readers will undoubtedly be content to learn that this is so. As long as they know that someone else has taken the trouble to gather the necessary figures, they'll be quite ready to take my word for it and will wait for me to let them know of the most important items. Others, however, will be suspicious that I'm keeping something up my sleeve. They'll not be content until they've at least glimpsed these mysterious tables themselves.

If you're in that latter category, you'll find the tables (in one form or another) in the back of many statistics textbooks. (They enable you to work out just what proportion of the total area under the curve lies between any two points along the base-line.) Alternatively, you can look at the panel on the opposite page, where I show a *selection* from such a table.

If, however, you are willing to take my word for the main proportions under the normal curve, simply read on. In this book, for our present purposes, the approximate proportions shown in the diagram below should suffice:

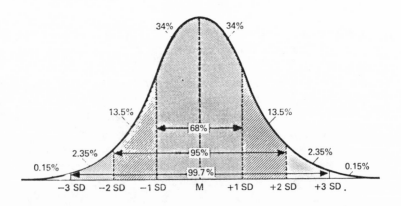

A selection of proportions under the normal curve

The figures below are a *selection* from a table showing the proportions under the normal curve. That is, they answer the question: what area lies between the mean and any point on either side of it, whose distance away is measured in units of the standard deviation?

The table below will allow you to see the proportion between the mean and, say, 1.25 SD or 1.50 SD. But the *full* table would show the proportion between the mean and 1.26 SD ... 1.265 SD ... and so on. However, I have picked out enough figures to show you the general pattern.

To take an obvious example: at 0 SD away from the mean (that is, at the mean itself), .5000 (or 50%) of the area lies beyond.

But we see that, between the mean and (±)1.00 SD. lies 0.3413 of the area (or 34.13%). This we see in column A. In column B we see that 0.1587 (or 15.87%) lies beyond 1.00 SD.

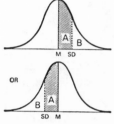

I have bracketed together the figures of (1.96 and 2.00 SD) and (2.50 and 2.58 SD). The reason is that, in this book, I am using 2 SD and 2½ SD as 'easy' approximations of the SD-values that cut off 95% and 99% of the area (which are actually the proportions cut off by 1.96 and 2.58 SD).

To find the proportion of the curve included on *both* sides of the mean, we have to *double* the appropriate figure. Thus 2 × 0.3413 = 0.6826 lies within 1 SD either side of the mean. And 2 × 0.1587 = 0.3174 lies beyond this range.

The full tables would allow you to work in *reverse* also. Suppose, for some reason, you wanted to know how far from the mean you had to travel (in terms of SD) to include, say, 49.7% of the area. You would look in the *body* of the table and (as you can see from the underlined figure alongside) read off the distance of 2.75 SD. The full tables would, of course, enable you to do this for *any* proportion: 20%, 25%, 75%, or whatever.

SD	A area between mean and SD	B area beyond SD
0.00	.0000	.5000
0.25	.0987	.4013
0.50	.1915	.3085
0.75	.2734	.2266
1.00	.3413	1587
1.25	.3944	.1056
1.50	.4332	.0668
1.75	.4599	.0401
1.96	.4750	.0233
2.00	.4772	.0228
2.25	.4878	.0122
2.50	.4938	.0062
2.58	.4951	.0049
2.75	.4970	.0030
3.00	.4987	.0013
3.25	.4994	.0006
3.50	.4998	.0002
4.00	.49997	.00003

As you'll remember, a normal curve can be tall and thin or wide and spreading – but its proportions will always be the same.

It may interest you to check how far these 'ideal' proportions are true of our not-so-'normal'-looking distribution of pulse-rates. There, you'll remember, we found a mean of 79.1 beats per minute and a standard deviation of 7.6 beats per minute, so:

$$M - 3\ SD = 79.1 - (3 \times 7.6) = \quad 56.3 \text{ beats per minute}$$
$$M - 2\ SD = 79.1 - (2 \times 7.6) = \quad 63.9 \text{ beats per minute}$$
$$M - 1\ SD = 79.1 - \qquad\ 7.6 = \quad 71.5 \text{ beats per minute}$$
$$M \qquad\quad = \quad 79.1 \text{ beats per minute}$$
$$M + 1\ SD = 79.1 + \qquad\ 7.6 = \quad 86.7 \text{ beats per minute}$$
$$M + 2\ SD = 79.1 + (2 \times 7.6) = \quad 94.3 \text{ beats per minute}$$
$$M + 3\ SD = 79.1 + (3 \times 7.6) = 101.9 \text{ beats per minute}$$

None of these seven figures was actually observed, of course, but let's put them into the distribution of recorded values and see into what proportions they split the observations.

56.3	62	**63.9**	64	65	66	68	70	71	71	**71.5**	72
72	73	74	74	75	75	76	77	77	77	78	78
78	79	79	79	**79.1**	80	80	80	80	81	81	81
81	82	82	82	83	83	85	85	86	**86.7**	87	87
88	89	90	90	92	94	**94.3**	96	**101.9**			

There are 50 observations (half of 100) altogether. So it's easy to work out what percentage of them lies between each pair of SD-points. We simply count them and double the numbers to get the percentages. For instance, there are 7 observations between –1 SD (71.5) and –2 SD (63.9), which makes 14%. I've marked it above the curve on the distribution below:

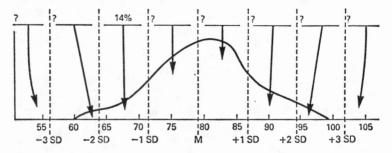

Now fill in the percentage of observations for each of the seven remaining areas on the diagram.

* * * * * * *

The percentages on your chart (from left of –3 SD to right of +3 SD) should read as follows:

Actual: 0%   2%   14%   34% 32% 16%   2%   0
(Predicted): 0·15% 2·35% 13·5% 34% 34% 13·5% 2·35% 0·15%

It seems that the proportions are reasonably close to those predicted by the normal curve.

Comparing values

So, when we know the mean and standard deviation of a bell-shaped distribution, we also know a lot about its internal proportions. One thing this enables us to do is to compare two or more observations from *different* distributions. For example, suppose we know that Bill scored 90 in the Economics exam and Linda scored 80 in Law. We know also that, as is often the case, the students' scores within each subject are approximately 'normal' in distribution. Now, which of our two students has the better score, compared with his or her fellow-students?

Many people would jump to the conclusion that it must be Bill because he has a higher score than Linda. But, conceivably, Bill's 90 might be the lowest score in Economics, while Linda's 80 might be the highest in Law. Next, we need to ask how the two scores relate to those of other students. Well, the mean score in Economics was 60, and in Law it was 65. So both scores are above average, Bill's by 30 points and Linda's by 15 points. Does that make Bill's the better score? Not quite. We need to know how many other students exceeded the mean by as much as Bill and Linda.

Suppose I now tell you that in Economics the standard deviation was 15 marks, while in Law it was 5 marks. Can you use this information (and what you know about the proportions of the normal curve) to decide: what proportion of the Economics students exceeded Bill's score of 90? And what proportion of the Law students exceeded Linda's score of 80? (Look at the diagram on page 77 to refresh your memory.) So which student had the better score, compared with his or her fellow-students? (It may help you to sketch out a diagram.)

*　*　*　*　*　*　*

In Economics, the mean was 60 and the standard deviation was 15 marks. So Bill's score of 90 was *two* standard deviations above the mean (2 × 15 marks above 60), and Bill's score would be exceeded by only 2.5% of his fellow-students. In Law, the mean was 65 and the standard deviation was 5 marks. So Linda's score of 80 was *three* standard deviations above the mean (3 × 5 marks above 65), and Linda's score would be exceeded by only 0.15% of her fellow-students. Clearly, then, as we can see in the diagram below, Linda has the better score, compared with fellow-students.

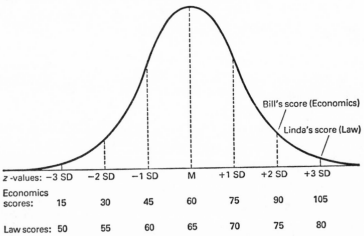

The above chart shows the base-line of the distribution measured off in units of the standard deviation – often called z-VALUES – and in the scores of the Economics and the Law distributions.

In effect, we have translated the observations of both distributions into z-values.* That is to say, we have re-expressed them as being so many SDs above or below the means of their distributions. Thus, we can see at a glance that an Economics score of 30 is equivalent to a Law score of 55, since both are two standard deviations below the mean.

Again, which is the better score, 60 in Economics or 60 in Law?

* * * * * * *

60 in Economics is the mean score but, in Law, 60 is one standard deviation below the mean. So 60 in Economics is the better score.

As you see, converting values to z-values enables us to compare two different distributions. It could even allow us to compare values from two different *variables*. For example, we might ask whether a particular student's blood-pressure was as high (relative to that of other students) as his pulse-rate. We'd be asking, 'Does his blood-pressure exceed the mean blood-pressure by as many standard deviation units as his pulse-rate exceeds the mean pulse-rate?'

So, if you know the mean and standard of a bell-shaped distribution, you can do a great deal. You can estimate what proportions of the observations will fall above or below or between various values of the variable. And you can compare values from two different distributions (or even two different variables) by first translating them into z-values. Furthermore, whether your distribution is bell-shaped or not, you can use it as a sample from which to make predictions about some larger population. How confident you can be about doing so is what we'll discuss in the next chapter.

*Any value in a distribution can be converted into a z-value by subtracting the mean of the distribution and dividing the difference by the standard deviation. For example, when the mean is 50 and the standard deviation is 10, a raw score of 60 would have a z-value of $\frac{60-50}{10} = \frac{10}{10} = 1$. That is, 1 SD above mean = z-value of 1.

5. From sample to population

We saw in Chapter 4 that if we know the mean and standard deviation of a sample, we can often form a pretty good idea of its overall shape. We can, for example, make a well informed guess as to how many observations were made between this and that value, or how many exceeded such-and-such a value. This can be useful if we are reading a report of someone else's investigation and have no access to the raw figures which the two key measures (mean and standard deviation) are meant to summarize.

However, it is usually not the shape of the sample itself that we are interested in – or, at least, not for its own sake. Rather, we are concerned with how far we can *generalize* from it. Is what we see in the sample true also of some wider population? If the mean pulse-rate among the 50 students tested is 79.1 beats per minute, would that be the mean also if we tested all the students in the college? The short answer, of course, is No. But (assuming the sample is random) the two means would be similar (as would the dispersion). *How* similar is what we'll need to think about next.

Estimates and inferences

The figures we use to describe our sample can be seen as ESTIMATES of the figures that would truly describe the population. In statistical theory, the distinction between sample and population is basic. It is underlined by calling the sample figures (for mean, mode, median, range, standard deviation, inter-quartile range, etc.) STATISTICS; while the true mean, mode, etc., of the population (which we cannot know for certain) are called PARAMETERS.*

Furthermore, each statistic is normally represented in formulae by a Roman letter (e.g. sample mean = $\bar{x}$) while parameters are represented by Greek letters (e.g. population mean = μ). Since we won't be using formulae, I'll not ask you to learn the necessary Greek. I'll simply abbreviate 'sample-mean' to 'S-mean' and 'population-mean' to 'P-mean', etc., where necessary. (By the way, to help your memory, notice that 'sample' and 'statistic,' which go together, both begin with S. Similarly, 'population' and 'parameter' both happen to begin with P.)

So, a statistic is used to estimate a parameter. Knowing the S-mean, we can estimate the P-mean; S-range suggests P-range; S-standard deviation gives some idea of P-standard deviation; and so on. The process involved is statistical *inference*. Let's see how it works.

Human nature being what it is, we don't need much information to prompt us into making inferences. Suppose you look up from this book now to find yourself being stared at by what you can only assume must be a gnome, pixie, or elf. He (or she or it) is wearing rather fanciful clothes and is about 10 centimeters in height.

This is the first gnome (or whatever) you've seen, and you've not heard reports of other people's sightings. A furtive glance around convinces you that there are no more lurking in the vicinity. So, here you have a sample of one. Not a lot to go on.

Still, what is your best guess as to the average height of the gnome population from which your sample of one is drawn? What further information would you need to be more confident about your estimate?

* * * * * * *

Your best guess as to P-mean would be 10 cm. However, since you've seen only one specimen, you have no idea how variable in

*This is also an 'in' word among politicians, journalists, and TV pundits. But they use it to mean something like 'dimensions' or 'constraints'; for example, 'We haven't yet faced up to the parameters of this situation.' They are not talking statistically.

height gnomes might be; and so you can't be sure how likely it is that you've been visited by a particularly small one or large one. So you need to see more.

All the same, it would be reasonable to assume that gnome heights are distributed according to the normal curve. So you'll realize that some distribution curves for the population are more likely than others. For instance, which of the curves illustrated below seem *less* likely and which *more* likely to describe the population? Why?

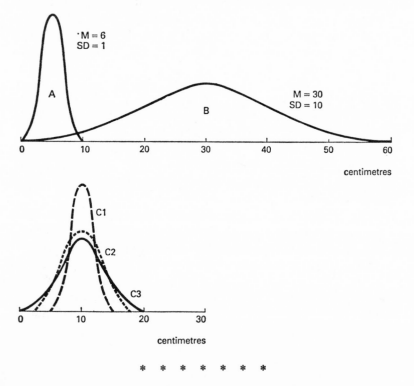

* * * * * * *

The curves A and B are least likely to describe the gnome population. Curve A would suggest that the gnome who is watching you is more than 3 SD above the mean which, as you'll remember, would make him one of the tallest 0.15%. Similarly, curve B would imply that he is more than 2 SD below the mean, and therefore one of the smallest $2\frac{1}{2}\%$. So B is a little more likely than A, but still very unlikely compared with *any* of the curves C1, C2 and C3. Each of these shows distributions with a mean of 10 cm; but, since we know nothing yet about variability in the gnome species, C1 indicates small dispersion, C2 medium, and C3 wide dispersion.

Clearly, as shown below, any of these C-curves (let's say C2) can be shifted to left or right to some extent and still remain fairly plausible. But the further it moves, putting 10 cm further and further from its centre, the less likely it is on the basis of the one specimen we've seen so far.

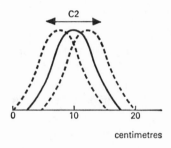

centimetres

But suppose our gnome is suddenly joined by four chums – measuring 9.2, 9.6, 10.3 and 10.5 cm. We can now make a new estimate of the population-mean (9.9 cm) based on a sample of five. We can also make an estimate of the variability or *dispersion* in the population based on that in the sample. In fact, the sample-standard deviation is only about 0.5 cm, suggesting few gnomes in the population are likely to be much smaller than $8\frac{1}{2}$ cm or bigger than $11\frac{1}{2}$ cm (that is, mean $\pm$ 3 SD). It begins to look as though C1 might be a somewhat better guess as to the shape and position of the population curve.

What is the point of all this speculation about the heights of gnomes? Simply to show that we can start making inferences in novel situations on the basis of very little information; but that we must modify those inferences as more information becomes available. With little information, many inferences are possible (though some are very unlikely). With increasing information, more and more of the possible inferences can be ruled out:

fewer and fewer of those that remain will fit the observed facts of the sample.

But we can never end up with certain knowledge about the population. Not unless we measure the whole population – in which case, we are not making an inference. The best we can expect to do is to say that the 'true' P-mean or P-standard deviation, or whatever, has such-and-such a probability of lying within such-and-such a range of possible values. Inference is inescapably attended by *error*, which all the formulae and computations of statistical method will not eliminate. All they can do is to make it explicit.

The logic of sampling

I am not going to bother you with those formulae and computations; but it is important for you to understand the logic that lies behind them. Let's return again to thinking about the pulse-rates of students in our college. Suppose we want to estimate the mean pulse-rate for the population as a whole. We take a random sample of 10 students and find that the S-mean is just over 78.6 beats per minute.

Now suppose we were to take four *more* random samples of 10 students each. Would you expect the sample-mean in *each* case to be 78.6 beats per minute?

* * * * * * *

If you expected the same mean each time, you'd almost certainly be disappointed. With four more samples of 10, we'd most probably get four *different* S-means. For instance, if we take each *row* of the table on page 42 as a random sample of 10, we get the following S-means: 78.6; 79.6: 78.0; 79.9; 79.4.

Clearly, dispersion or variability is not a characteristic simply of the observed values *within* a sample. The *means* of samples *are variable too*. So, also, are their other statistics. This variability from one sample to another is known as SAMPLING VARIATION.

Now if we pool our five samples, we get a mean for the total sample of 50 students of 79.1 beats per minute. (This is the mean of the five means.) But another random sample of 50 students would give yet a different mean – maybe 79.4 or 78.9. If we took bigger samples (e.g. of 100 students at a time), their means would be less variable. But the dispersion (due to sampling variation) would never disappear altogether. So how do we cope with it?

A distribution of sample-means

What we do is this: think of taking a large number of random samples of a given size from the same population. Each of these many samples will have its own sample-mean. Now let's look at all those sample-means *on their own*. Let's see how they compare in size from smallest to largest, and how many we have of each size. In other words, imagine we make a *frequency distribution* of the sample-means themselves. (Are you still with me? We're treating 'size of sample-mean' as the variable here.) So we get a DISTRIBUTION OF S-MEANS.

This distribution of S-means will itself have a mean – the mean of the S-means. If we have taken a large enough number of samples, the distribution curve of their S-means will center around the *population-mean*. That is, the *mean* of the S-means will be equal to P-mean.

What *shape* do you imagine the distribution curve of S-means might have?

*   *   *   *   *   *   *

As you may have guessed, the distribution of sample-means will approximate to the *normal* distribution. Why is this? Well, in the population, values become increasingly scarce as they get bigger or smaller than the mean of the population. So, in sampling from that population we are likely to get more observed values similar in size to the P-mean than very different from it. Hence, samples whose observed values are mostly close to the P-mean are likely to be more numerous than samples with a lot of values very distant from P-mean. Thus, in turn, samples whose S-means are similar to the population-mean are also likely to be more frequent than samples whose S-means are very different from it.

So, as you see in the diagram below, if we were to make a distribution of S-means, we would find that it centers around P-mean, with fewer and fewer larger and smaller S-means either side of it:

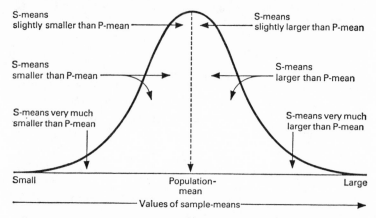

This is perhaps the most elegant and valuable feature of the normal distribution: *whether or not* the population (and therefore the samples) is normally distributed, the means we compute from samples *will* be approximately normal in distribution. The bigger the samples (and the more bell-shaped they are), the closer to normal will be the distribution of their S-means.

In the real world, however, we never see a distribution of sample-means, from which to estimate the population-mean. Usually, as you'll remember, we have to make an estimate of P-mean on the basis of just *one* sample. We shall certainly be somewhat in error if we use this single sample-mean as an estimate of P-mean. But . . . shall we be *more likely* to make (a) a big error? or (b) a small error? (Check with the diagram above.)

* * * * * * *

We are more likely to make a *small* error. In estimating P-mean from the S-mean of a single sample, the *bigger* the mistake we could make (that is, the bigger the difference between S-mean and P-mean), the *less* likely we are to make it. As you can see in the diagram above, the bigger differences between S-mean and P-mean are out towards the tails of the distribution. In other words, fewer and fewer samples have means that are so much smaller or larger than that of the population.

Now let's look at the amount of *dispersion* among the sizes of the sample-means. Just as the distribution of S-means has its *own* mean (equal to P-mean), so it has its own standard deviation.

How will this standard deviation among the means of samples compare with the standard deviation of the population itself? Would you expect it to be larger, smaller, or about the same? (You may find it helpful to sketch out distribution curves in coming to your decision.)

* * * * * * *

The standard deviation of the distribution of sample-means will be *smaller* than that within the population. That is to say, the means of samples will not differ among themselves as much as do the raw values in the population. The distribution of sample-means will then, in general, be less dispersed than are individual samples.

We can illustrate these relationships on the diagram below. It has three curves, representing (1) the distribution of the population, (2) the distribution of just one (rather large) sample, and (3) the distribution of means from a large number of samples.

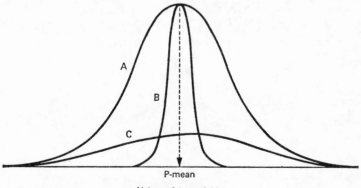

P-mean

Values of the variable

Which curve is which?

* * * * * * *

The population distribution is (A). The sample, with its mean slightly to the right of the population mean, but with similar range, is (C). And the distribution of sample means, with the same mean as the population but a considerably smaller dispersion than either the sample or the population, is (B).

So, even though we take only one sample, and therefore have only one sample-mean, we can think of it as belonging to a distribution of *possible* sample-means. And, provided we're thinking of samples of reasonable size, this distribution will be normal. When we imagine such a distribution for a statistic (like the mean or standard deviation) drawn from samples, we call it a SAMPLING DISTRIBUTION OF A STATISTIC.

As we've seen, the sampling distribution of a statistic will have a mean and standard deviation of its own. In the case of the sampling distribution of S-means, the mean will be the same as that of the population; but the standard deviation will be smaller than that of the population.

The standard deviation of a sampling distribution (e.g. of the sample-means) is called the STANDARD ERROR (SE). (This is to distinguish it from the standard deviation of a sample or of the population.) The idea of standard error enables us to calculate the chances that a particular sample-mean will be much bigger or smaller than the population-mean.

For instance, since the distribution of sample-means is normal, we can say, for example, that about 68% of all sample-means will lie between the mean and 'one standard deviation *of the sample-means*' (one standard error) either side:

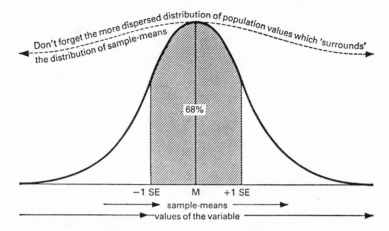

Suppose we knew that the mean of the sampling distribution of mean pulse-rates was 78 beats per minute with a standard error of 1.4 beats per minute. In what percentage of samples would we expect to find a mean *bigger* than 79.4 beats per minute?

* * * * * * *

A sample with a mean of more than 79.4 beats per minute would occur in 16% of samples. That is, 79.4 is 1 standard error (SD of the S-means) above the mean (78); 50% of samples will have a mean less than 78; 34% will have a mean between 78 and 79.4; samples with a mean exceeding 79.4 will therefore be in the remaining $100 - (50 + 34)\% = 16\%$. For confirmation, see the shaded area under the curve below:

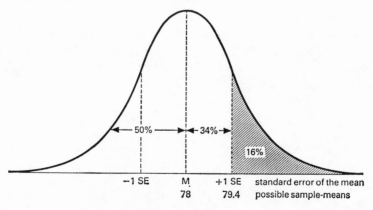

This is all very well, you may say, but how can we expect to know the mean and standard deviation of a sampling distribution of means – consisting of the sample-means from a huge number of samples? All we usually have is one sample. We know its mean and the standard deviation of the observed values within it. How are we to determine the standard error of *all* the possible sample-means when we have only *one* sample-mean to go on?

In fact, it can be shown that the standard error of the mean is related to three factors: (1) the standard deviation within the sample, (2) the size of the sample, and (3) the proportion of the population covered by the sample.

Let's think about these factors one by one. (1) Would you expect the standard error of the means to be larger when the samples concerned have small standard deviations or when they have large standard deviations?

* * * * * * *

(1) The more variable the values within a sample, the more likely it is that the means of such samples would vary greatly. So a bigger sample-SD will lead to a bigger SE of the means from such samples. (Actually, SE depends on variability in the population; but provided the sample contains 30 or more observations, sample-SD is a sufficiently accurate estimate of population-SD.)

Now for the second factor. How does the size of SE (the variability of sample-means) relate to the size of the samples concerned? (2) Would you expect the standard error of the mean to be larger when samples are small or when they are large?

* * * * * * *

(2) The larger the sample, the closer (on average) would the means of such samples be to P-mean in size. Therefore, the less would be the standard error. The diagram below shows how the dispersion of a sampling distribution of the sample-mean is smaller with big samples than it is with small ones:

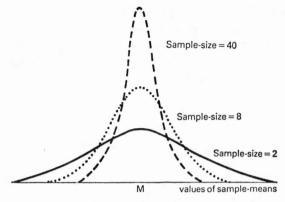

Again, I'll just remind you that sampling distributions such as these are the imaginary result of taking a huge number (actually an infinite number) of samples of the relevant size.

Now for the third factor influencing the size of SE. (3) How would you expect the size of the standard error to be influenced by increasing the *proportion of the population* covered in samples?

*   *   *   *   *   *   *

(3) The larger the proportion covered in samples, the lower the variability (SE) of the means. But, and it is a big *but* – this factor is remarkably *unimportant*. It has some, but very little, influence on the size of the standard error. It is the size of the sample itself, the sheer *amount* of information available (not the percentage of possible information), that determines the accuracy of the results. Some extra accuracy is added (that is, the standard error will be reduced) if the sampling fraction (the proportion of the population covered) exceeds 10% – but not much. This is just as well because real-life samples are often well below 10% – especially, of course, where the population is infinite (like all white mice) or expensive to sample from (say, the testing 'to destruction' of light bulbs).

So for practical purposes we can say that the standard error of the mean depends on the size of the sample and its standard deviation. Standard error will be smaller than the sample-SD. How much smaller depends on the size of the sample. The larger the sample, the greater the proportion by which standard error is smaller than the standard deviation.

In fact, the standard error of the mean is calculated by dividing sample-SD by the *square root* of the number of observations in the sample. Thus a sample of 100 examination scores with a standard deviation of, say, 15 marks would lead us to expect the means of all such samples to have a standard error of:

$$\frac{15}{\sqrt{100}} = \frac{15}{10} = 1.5 \text{ marks} = \text{SE}$$

To halve the size of the standard error we'd need to quadruple the sample size (from 100 to 400):

$$\frac{15}{\sqrt{400}} = \frac{15}{20} = 0.75 \text{ marks} = \text{SE}$$

The smaller the standard error, the more confident we can be that our sample-mean is close to the population-mean. But big increases in sample size produce relatively small improvements in accuracy.

It seems to fly in the face of common sense that the size of the population has little effect on the accuracy (the standard error) of the sample. Nevertheless it is so. If the sample is big enough to contain adequate information, don't worry about how small it is compared with the population. How to decide whether it *is* big enough, we'll think about towards the end of this chapter.

Estimating the population-mean

So the standard error of the mean is, in effect, determined by the size and standard deviation* of the sample. We can now say,

for example, that the range: P-mean $\pm 1 \sqrt{\dfrac{\text{sample-SD}}{\text{sample size}}}$

that is, P-mean ± 1 SE

*Strictly speaking, SE should be determined by the standard deviation of the population rather than of the sample. But, with samples of about 30 or more, it is safe to use S-SD as an estimate of P-SD.

would contain about 68% of all sample means. But what use is
this? All we have is *one* sample-mean. How do we use it to make
an inference about the size of the population mean?

Here is how: look at the diagram below, which shows a
distribution of sample-means. The distance of 1 SE is marked off
on either side of the P-mean. (68% of all sample-means fall
within this range.)

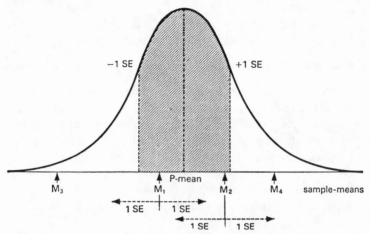

Think about all those S-means in the range P-mean ± 1 SE.
(I've indicated two 'specimens,' M₁ and M₂.) For every such
S-mean, the range S-mean ± 1 SE will contain the population
mean. That is, the population-mean will lie within one SE either
side of any such sample-mean. (I have drawn lines either side of
M₁ and M₂, equal in length to 1 SE, and you'll notice that, in
each case, one of the lines crosses the mid-point of the distribution, which is P-mean.)

What about sample-means like M₃ and M₄? Would the range
S-mean ± 1 SE contain the population-mean in their cases also?
(Draw lines of 1 SE in length on either side of M₃ and M₄ if
you wish to check.)

* * * * * * *

As you'll see in the diagram above, M₃ and M₄ lie *more* than 1 SE
away from the population-mean. So a range of 1 SE either side
of such sample-means would *not* contain P-mean.

So long as an S-mean is no more than 1 SE away from P-mean, the range S-mean $\pm$ 1 SE will contain the P-mean. Now, we know that about 68% of all sample-means (including M_1 and M_2 above) are no more than 1 SE away from the P-mean. Therefore, whatever the mean of any sample we happen to have taken, there is a 68% probability that the range S-mean $\pm$ 1 SE contains the population-mean – and, of course, a 32% probability that it does not.

Consider our random sample of 100 students' examination marks: the mean mark is 50 and the standard deviation is 15. So the standard error of the mean is $\dfrac{15}{\sqrt{100}} = \dfrac{15}{10} = 1.5$ marks. How would this help us estimate the mean mark of *all* students who took that exam? Well, there is a 68% probability that the P-mean lies within 1 SE of the S-mean. Thus, we can say that we are 68% confident that the mean mark of the total population of students lies between S-mean $\pm$ 1 SE = 50 $\pm$ 1.5 marks, which gives a range of 48.5 to 51.5 marks. This range (or interval) would be called the 68% CONFIDENCE INTERVAL.

What is the probability that the true mean lies *outside* that rather narrow range of marks?

* * * * * * *

The probability that the true mean of the population lies outside the range is 32%. Since 48.5 to 51.5 marks is the 68% confidence interval, we can be equally confident that 32% of all sample means will *not* fall within it.

So our sample-mean does allow us to infer a population-mean of 50 $\pm$ 1.5 marks – but with only a 68% probability of being right. If we want to be more confident that our estimate is correct, we'll have to *widen* the confidence interval. For instance, we could be considerably more confident that the population-mean lies between 50 $\pm$ 3 marks. To increase our confidence in our estimate, we have here widened the range (the confidence interval) to 2 SEs either side of the sample-mean.

If we wanted to be practically 100% confident (99.7% anyway), we'd say the population-mean lay between sample-mean ± *how many* SE? (Check back to the diagram on page 77 if you need a hint.)

* * * * * * *

The range of S-mean ± 3 SD should contain practically all the sample-means (99.7% in fact). It is the 99.7% confidence interval. So, if our sample of 100 students' marks gave a mean of 50 and a standard error (not SD, remember) of 1.5 marks, we could be 99.7% certain that the true population-mean lay between $50 \pm 3 (1.5) = 50 \pm 4.5 = 45.5$ to 54.5 marks.

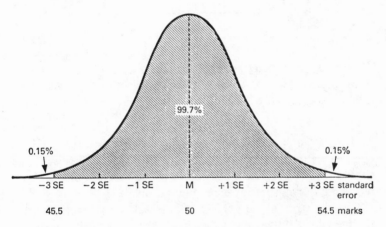

BUT, it could still be that 50 was one of the 0.3% of sample-means that lay outside that range. The probability is small but, as Aristotle once remarked, even the improbable does sometimes happen. In this context, it would happen three times in every 1,000 samples.

So we can make a wide estimate with a high degree of confidence or more precise estimates with lower degrees of confidence. In fact, the two confidence intervals most commonly used are those of 95% and 99%.

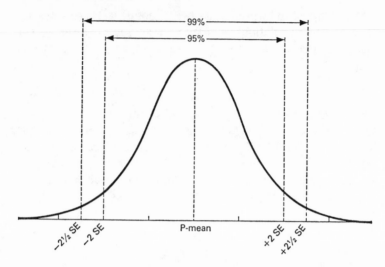

According to the diagram above (i) the 95% confidence interval is given by S-mean ± *how many* SE; and (ii) the 99% confidence interval is given by S-mean ± *how many* SE?

* * * * * * *

(i) The 95% confidence interval is S-mean ± 2 SE. (ii) The 99% confidence interval is S-mean ± 2½ SE. (More precisely, we should say 1.96 SE and 2.58 SE, but the rounded numbers above will serve our purposes well enough in this book.)

Let's illustrate these confidence intervals by returning to our sample of 50 pulse-rates which had a mean of 79.1 beats per minute and a standard deviation of 7.6 beats per minute. What is the true mean of the population (of nearly 1,000 students)? First we calculate the standard error of means from such samples:

$$\text{SE} = \frac{7.6}{\sqrt{50}} = \frac{7.6}{7.07} = 1.1 \text{ beats per minute.}$$

So we can be 95% certain that the true mean lies within the confidence interval: S-mean ± 2 SE

$$= 79.1 \pm 2(1.1)$$
$$= 79.1 \pm 2.2$$
$$= 76.9 \text{ to } 81.3 \text{ beats per minute.}$$

In other words, there is only one chance in twenty (5%) that the true mean of the population is less than 76.9 or more than 81.3 beats per minute.

If we felt that 95% certainty was not enough and we wanted to be even more confident that the range we quoted included the true population-mean, we might use the 99% confidence interval:

$$79.1 \pm 2\tfrac{1}{2}\,(1.1)$$
$$= 76.35 \text{ to } 81.85 \text{ beats per minute.}$$

There is only one chance in a hundred that the true mean falls outside this range. But remember, if one hundred such samples were taken, probability warns us to expect one of them to have a mean from which we would wrongly infer a population-mean outside of this range.

Estimating other parameters

Other parameters of the population can be estimated from sample statistics in much the same way. One parameter we might particularly wish to know is a PROPORTION. This will be the case when the characteristic we are interested in is a *category*-variable, say, whether or not a graduate is unemployed.

For instance, we may be anxious to know what proportion of students leaving colleges last year are still unemployed. The proportion in the whole population can be estimated from the sample – but with a range of possible error indicated.

Again, the underlying idea is that, in a large number of samples, the proportion of unemployed would vary from one to another – 0.19 in one, 0.23 in another, 0.21 in a third, and so on. But the distribution of such samples would be approximately normal and centered around the true proportion (perhaps 0.22 of last year's graduates are unemployed). We can thus calculate the STANDARD ERROR OF THE PROPORTION. However, since a single sample will have no variability from which to estimate variability across all samples, we use a different method of deciding standard error. It is calculated from the sample by multiplying the proportion we are interested in by the proportion remaining, dividing by the number of cases in the sample, and taking the square root.

Thus, if a random sample of 100 ex-students shows that 20 of them are unemployed (and 80 are not):

$$\text{SE proportion} = \sqrt{\frac{0.2 \times 0.8}{100}}$$
$$= \sqrt{\frac{0.16}{100}}$$
$$= \sqrt{0.0016}$$
$$= 0.04$$

So, if we wanted to be 99% confident of the proportion of unemployed among all the students who left college last year, we'd say the proportion lay between: $0.2 \pm 2\frac{1}{2}(0.04) = 0.10$ to 0.30.

So, as you see, even the 99% confidence interval is very wide with a proportion. (It would have been even wider, as you can check yourself if you like arithmetic, had the proportion of unemployed in the sample been nearer 50%.) We can say with 99% certainty only that the number of unemployed ex-students in the population of, say, 5,000 ex-students lies somewhere between 500 and 1,500. This is quite a range of error and perhaps too wide to be of any practical value.

Unfortunately, there is only one way we could reduce the width of that confidence interval (and still be 99% certain). Can you guess how we'd do it? (Check back to page 91 if you want a clue.)

*   *   *   *   *   *   *

The only way we could narrow the range is by taking a bigger sample. But it would have to be a much bigger sample to have real effect. Since the standard error is influenced by only the square root of sample size, a fourfold increase in the sample size would only halve the confidence interval. Thus, a sample size of 400 (in which 20% were unemployed) would lead us to expect between 750 and 1,250 unemployed ex-students among the population of 5,000.

Before carrying out an inquiry (whether into a category-variable or a quantity-variable), it can be worth conducting a 'pilot inquiry' to estimate the likely standard error. If, with a category-variable, the proportion approaches 50%, or if, with a quantity-variable, the standard deviation seems large, then you'll need a bigger sample than otherwise.

Thus, a metallurgist might be happy to infer the breaking point of a new alloy on the basis of a handful of specimens – because they show little variability. However, the psychologist wishing to generalize about the breaking point of human beings (e.g. in their reaction to stress) might be unwilling to do so except after observations of several hundred – because human beings are far more various in their behavior than are lumps of metal.

In general, if we want to be sure of estimating the mean or the proportion within the population with a certain precision (for example, ± 5 or $\pm 5\%$), and with a certain confidence level (for example, 99%), we could work out how large a sample we'd need to reduce the standard error to the necessary level. Whether we could afford to gather a sample of such size would, of course, be another matter. If the cost or danger of making incorrect inferences would be high (say, in testing a potentially harmful drug), we might just feel compelled to afford it.

To sum up this chapter: whatever our sample and the statistics we draw from that sample, we can never use them to say exactly what the mean or proportion (or any other parameter) of the population really is. The bigger the sample and the less variable its observations (or the more unevenly divided between one category and another), the more certain we can be. But we can never pin down one figure with 100% certainty. Our best estimate of the mean (or proportion) must have a range of uncertainty around it. And it must be stated in terms of probability: there is an X% likelihood that the true mean (or proportion, etc.) falls within such-and-such a range. Not very satisfactory, perhaps, but the best we can do in the real world!

6. Comparing samples

In Chapter 5 we considered how to make inferences about a population based on what we know of a single sample. We can now consider another highly important area of statistical inference: looking at two (or more) different samples and asking whether or not they imply a real difference in populations. This is what lies behind questions like: 'Are girls more intelligent than boys?' 'Does treatment with the new drug help more patients than would the standard treatment?' 'Which of these four methods is the most effective way of teaching French?' 'Are the costs of running model-X car more variable than those of running model-Y?'

From the same or different populations?

As an example, suppose we measure the blood-pressures of two random samples of students, 50 males and 50 females. What can these two random samples tell us about the *difference* between the blood-pressures of male and female students *in general*? Are the samples so similar that we'd be justified in lumping them together and saying they are from the same population; or are they so different that they signify two different populations? (For instance, in order to *estimate* a student's blood pressure, would it help to know that student's sex?)

Just suppose the distribution curves for our two samples looked something like this:

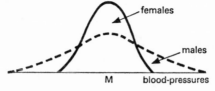

Would we be justified in thinking that both samples came from the same population? That is, would it seem likely that the population of female blood-pressures has the same mean and standard deviation as the population of male blood-pressures? Why, or why not?

* * * * * * *

Although the means are much the same in the two samples above, the standard deviations are very different. We know that the standard deviation of a sample (provided it has 30 or more members) can be taken as a reasonable estimate of the standard deviation in the population from which that sample comes. So, having two very different standard deviations, the samples pictured above would seem to come from different populations. The population of male blood-pressures would appear to be more variable than that of female blood-pressures.

Another, more difficult comparison arises, however, when the standard deviation is much the same in each sample, but the *means* differ. Look at the diagram below: it shows three different

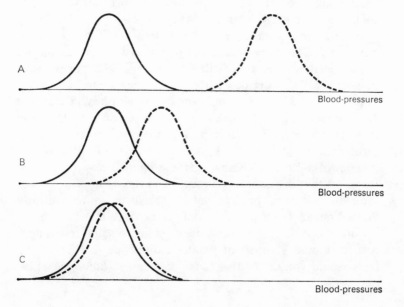

pairs of sample distributions. In each pair, the dispersion (e.g. standard deviation) of each sample is the same. But the means differ between the two samples in each pair.

In which of these cases (A, B or C) would you feel (i) most confident and (ii) least confident that the pair of samples came from *different* populations?

* * * * * * *

In each of the three pairs of samples, the means are different. In (A) they are very different, in (B) they are much less different, and in (C) they differ scarcely at all. Now, we know about the chance factors of sampling variation, so we don't expect two random samples drawn from the same population to have exactly the same mean. But we also know that we are more likely to draw a pair of samples with means fairly close together than a pair whose means are far apart. (Do you remember the distribution of possible sample-means on page 88?) Therefore, our chances of choosing pair (A), above, from a single population are small compared with our chances of choosing pair (B), and especially small compared with our chances of choosing pair (C). So, to look at it the other way round, (A) is the pair of samples I'd feel *most* sure came from *different* populations. Similarly, (C) is the pair I'd feel *least* sure came from populations with different means.

So, given two samples with similar dispersion but different means, how do we decide whether or not they are from the same population? Let's work it out, using our blood-pressure example: two random samples, 50 males and 50 females. Suppose the mean male blood-pressure is 120 mm, and the standard error of the mean (calculated from the standard deviation of all 50 blood-pressures) is 1.6 mm. Knowing these figures, we can be 99% confident that the population mean (and the mean of any other sample of 50 from that population) would lie within a clearly defined range. There would be only one chance in a hundred of getting a sample of 50 with a mean falling outside this range. Just to remind yourself of what we discussed in Chapter 5: How would you define this range: S-mean ± *how many* SE?

* * * * * * *

We could be 99% confident that the population-mean lies in the range: S-mean $\pm$ $2\frac{1}{2}$ SE (or 2.58 SE, to be more precise). So the population-mean lies in the range: $120 \pm 2\frac{1}{2}(1.6)$ mm. (If you're a non-mathematician, as I'm assuming, you won't be offended by my mixing vulgar fractions and decimals in this way!) So, for the male blood-pressures, P-mean lies between 116 and 124 mm.

Similarly, if the sample of women's blood-pressures had a mean of 110 mm and, let's say, the same standard deviation and therefore the same standard error of the mean, we could work out that there was only one chance in a hundred of finding such a sample with a *mean* falling outside the range 106 to 114 mm; that is, S-mean $\pm$ $2\frac{1}{2}$ (1.6).

Let's put all these figures in a table, just to keep check on how far we've come:

Sample	S-mean	Estimate of P-mean (99% level)
Males	120 mm	116 to 124 mm
Females	110 mm	106 to 114 mm

What can you conclude from this so far?:

(a) no woman will have higher blood-pressure than a man? or
(b) very few women will have higher blood-pressure than a man? or
(c) no random samples of women will have a higher mean blood-pressure than a similar sample of men? or
(d) very few random samples of women will have a higher mean blood pressure than a similar sample of men?

* * * * * * *

Although there will be a considerable number of women who have a higher blood-pressure than men, very few samples of fifty women (d) will have a higher *mean* blood-pressure than fifty men.

See how this looks on a diagram. (Strictly speaking, the curves for the sampling distributions of the means should be infinitely tall if they are to appear on the same chart as the sample-curves.

After all, the area under each sample-curve is supposed to represent only 50 observations, while the sampling distributions of the means are supposed to represent populations of infinite size. However, although they're not tall enough on this scale, the lower parts of those two curves are correctly shown; and it is the overlap at the base that we are chiefly interested in here.)

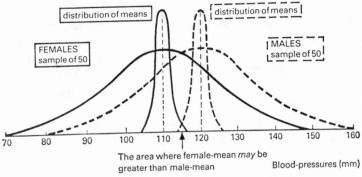

The area where female-mean *may* be
greater than male-mean   Blood-pressures (mm)

While there is a considerable overlap between the two samples, there is scarcely any overlap between the two distributions of *possible sample-means*.

We can calculate the probability of picking out a pair of samples (one female, one male) in which the female mean equalled or exceeded the male. In one in 100 cases, we might expect the men's mean to be outside the range 116 to 124. And in one in 100 cases, the women's mean might be outside the range 106 to 114. In only one out of 100 × 100 pairs of samples of fifty would both means be outside the limits. So even that possibility is extremely unlikely – one in 10,000 chances. But, even then, the female mean might be low while the male is high; or both may be low; or both may be high; only when the female mean is high and the male mean low is there a chance that the female mean would exceed the male. We would therefore have to take 4 × 10,000 samples to cover all these four equally possible eventualities. So the probability of picking out a pair of samples of 50 in which the female mean exceeds the male mean is less than $\frac{1}{40,000}$. It is so small as to be negligible.

The sample statistics (mean, and standard deviation, and standard error) have signified a difference in the populations from which the samples were drawn. We can be certain that the female blood-pressures are, statistically speaking, drawn from a different population than the male blood-pressures. There is a reliable and 'systematic' difference between male and female blood-pressure at this age.

Significance testing

When we compare samples in this way we are applying a TEST OF SIGNIFICANCE. We are asking whether the difference between samples is big enough to signify a real difference between populations. (Whether such a difference, even if real, is in any sense important, is a matter we'll consider later.)

How exactly do we carry out a significance test? I've already illustrated one way, based on the likelihood of an overlap between the distributions of sample means. A more common approach, however, is to think in terms of a DISTRIBUTION OF DIFFERENCES between means. At first, this idea may strain your imagination even more than did that of a distribution of sample-means! But you'll soon find it makes sense.

Suppose we consider a population (e.g. the blood-pressures of male students) and draw samples of, say, 100 from it. Let's say we draw two random samples at a time – call them A and B. Naturally, we would not rely on the mean of sample A being the same as the mean of sample B. Most often, perhaps, they would be the same; but quite often mean-A would be slightly bigger than mean-B, or vice versa; and occasionally, mean-A would be considerably bigger than mean-B, or vice versa. If we compared an infinite number of pairs of reasonably large samples from this population, we could form a frequency distribution of the *differences* between pairs of sample-*means*. In other words, we are asking: With what frequency does mean-A considerably exceed mean-B? How often does it slightly exceed it? How often are the two means equal? How often does mean-B slightly exceed mean-A? How often does it exceed considerably? And so on.

This is to say, a *difference* between means is another statistic – this time descriptive of *two* samples, rather than one. As with other statistics, we can thus start thinking about its 'sampling distribution.' What would be the distribution of this statistic in an infinite number of such pairs of equally sized samples? Can you imagine what shape this distribution would have? Can you guess what its mean would be?

*   *   *   *   *   *   *

The distribution of *differences* between the means of pairs of samples from the same population would be approximately *normal*. (This would be so even if the distribution of the population itself were not normal.) The mean of the distribution would be *zero* – overall, the A-means would exceed the B-means as often and by the same amount as the B-means exceeded the A-means; small differences would be more frequent than big ones. Here is the distribution curve:

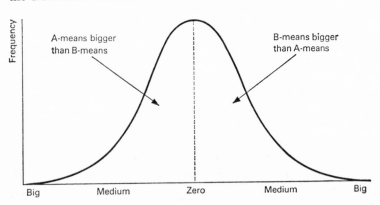

Size of difference between pairs of sample-means

This distribution curve takes a little getting used to. Although it looks like any other normal curve, it has one very distinctive feature of its own: the base-line does not simply, as with previous curves, represent larger and larger quantities as we read from left to right. Instead, the quantities (differences between means) *start large* on the left. Then, as we read towards the right, they diminish to zero (in the middle) and *then* start to increase until the maximum on the extreme right matches that on the extreme left. (You may wonder why we don't simplify things by laying the left-hand side on top of the right and have a skewed curve with a normal base-line running from 0 to maximum. But you'll see later that we need to identify the *direction* of a difference – which mean is bigger than which – as well as the size.) The center-point of the distribution (the mean) represents the frequency of pairs of samples with zero difference between their means. (50% of pairs should have mean-A bigger than mean-B; while in the other 50% of sample pairs, mean-B should be bigger than mean-A.)

Suppose we take three pairs of samples from this population. The means (in mm) happen to be as follows:

(i)   sample A-mean = 118, sample B-mean = 120;
(ii) sample A-mean = 121, sample B-mean = 120;
(iii) sample A-mean = 121, sample B-mean = 122.

How big is the *difference* between the means of the two samples in each case? Will each difference belong on the *left-* or the *right*-hand side of the distribution? Which of the three differences will be *furthest* from the center (the mean) of the distribution?

*   *   *   *   *   *   *

In (i), sample B's mean is 2 mm *bigger* than sample A's, so it belongs on the right-hand side of the distribution. In (ii), sample A's mean is 1 mm bigger than the B-mean; so that difference belongs on the left-hand side. In (iii), sample B's mean is 1 mm bigger than sample A's mean, so that difference belongs on the right-hand side. The biggest of the three differences between the sample-means is in pair (i), so that is the difference that will be furthest from the mean in the distribution.

This, then, is the sort of theoretical distribution we'd expect from the *differences* between the two means of a huge number of pairs of samples drawn from the *same* population. It is the

'SAMPLING DISTRIBUTION' OF THE DIFFERENCES between means. The mean would be zero, and the dispersion would depend on the amount of dispersion in the population. That is, the greater the variability within the population, the more variable the samples would be and the greater the likelihood of widely differing means.

The dispersion in the distribution of differences between sample-means can be measured in standard deviation units. As with the distribution of sample-means, this standard deviation is known as the standard error – this time the STANDARD ERROR OF THE DIFFERENCES BETWEEN MEANS (SE-diff). Hence, all the proportions of the normal curve can be assumed; that is, about 68% of all differences between sample-means will lie between 0 ± 1 SE-diff; about 95% will lie between 0 ± 2 SE-diff, and so on.

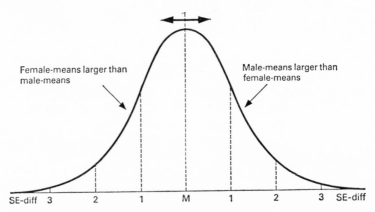

We'd expect practically all (99.7%) differences between the means of pairs of samples to be less than *how many* times the SE-diff?

*   *   *   *   *   *   *

Very few (0.3%) of differences will be bigger than $3 \times$ SE-diff.

Now, let's see how all this helps in an actual situation. How would it apply in comparing our two samples of blood-pressures? Right from the start, you'll remember, we noticed that the mean blood-pressure of the 50 men students was 10 mm higher than that of the 50 women. What we wanted to know was whether this difference was significant. That is, was it big enough to signify (or indicate) a difference in blood-pressure between such men students and women students in general? Or is this difference between the sample-means just one that we'd anyway expect to arise quite often by chance? To decide this we carry out a test of significance.

Now scientists are expected to be cautious – especially about evidence that might seem to support their theories. (As Charles Darwin once said: 'I always make special notes about evidence that contradicts me: supportive evidence I can remember without trying!') So statisticians have developed an especially cautious convention. The convention is that we test for a significant difference by asking: 'Could it be that the difference is *not* significant?'

So we start out as if assuming that there is *no real difference* between the blood-pressures of men and women. We assume, in effect, that they are all from the same population. Thus, the difference between the means of our two samples would be just one possible difference from the theoretical distribution above, where the mean difference is zero. This assumption, or hypothesis, is called the NULL HYPOTHESIS. (It attempts to nullify the difference between the two sample-means by suggesting it is of no statistical significance.)

From this point on, the null hypothesis is 'under assault.' If the difference between the two sample-means turns out to be too big to be explained away as the kind of variation that would often occur by chance between random samples, then we must reject it. It will not explain our findings. We shall have to replace it with an *alternative* hypothesis. The most usual alternative hypothesis would be simply that the two population-means are *not* equal.

This is not the only possible alternative hypothesis, of course. Can you think of another, perhaps one that would be more precise?

*   *   *   *   *   *   *

Another alternative hypothesis might be that the mean male blood-pressure is higher than the female mean, or vice versa. (You might even hazard that one was higher than the other by at least some specified amount, e.g. 5 mm.)

Anyway, the null hypothesis (that there is no real difference between the P-means) is assumed true *until* – and *unless* – the difference between S-means proves *too big* for us to believe easily that both samples could have come from the same population. So we are assuming that the difference of 10 mm between the means of the two samples of blood-pressures is just one difference from a normal distribution of differences between the means of such samples.

The mean of the distribution is zero. What is the standard deviation – the standard error of the mean of the difference? Of course, without taking an infinite number of pairs of samples, we can't know for certain. So we have to *estimate* the standard error of the difference between sample-means (just as we estimated the standard error of a mean). In fact, the standard error of the difference between means is found by *combining* the standard errors of the two means.

In the case of our two blood-pressure samples, the standard deviation of the fifty values in each sample happens to be the same (11.3 mm). Therefore the standard error of the mean in each case is also the same: $\frac{11.3}{\sqrt{50}} = 1.6$ mm.

In fact, we square the two standard errors, add them together and take the square root (so the SE-diff must always be larger than either of the SE-means):

$$\begin{aligned} \text{SE-diff} &= \sqrt{1.6^2 + 1.6^2} \\ &= \sqrt{2.56 + 2.56} \\ &= \sqrt{5.12} \\ &= \quad 2.26 \text{ mm} \end{aligned}$$

Following the null hypothesis (i.e. assuming these two samples come from the same population), then a distribution of the differences between the means of pairs of such samples would have a mean of zero and, apparently, a standard deviation (SE-diff) of about $2\frac{1}{4}$ mm:

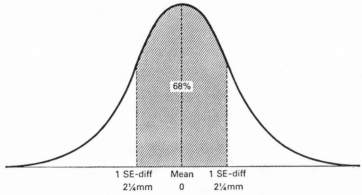

1 SE-diff   Mean   1 SE-diff
$2\frac{1}{4}$mm   0 $2\frac{1}{4}$mm

Difference between pairs of sample-means (mm)

So, *even if there were no real difference* between male and female blood-pressures, we could expect that in a hundred pairs of samples, 68 would differ in mean by up to $2\frac{1}{4}$ mm. Sometimes the male means would be up to $2\frac{1}{4}$ mm bigger than the female, sometimes vice versa. But such a difference, one way or the other, would have a probability of $\frac{68}{100}$ (that is, 68%). If this had been the difference in means between our two samples of blood-pressures we'd certainly have assumed it was just the result of random, chance variations in sampling – far too likely an occurrence to signify a real difference in the population means.

What would be the probability of the means of two randomly chosen samples differing by *more than* $2\frac{1}{4}$ mm? (Check with the diagram above.)

*   *   *   *   *   *   *

Differences of more than $2\frac{1}{4}$ are in the shaded *tails* of the distribution *below*. Since the middle section (0 ± 1 SE-diff) accounts for 68% of all differences between sample-means, the other 32% are shared between the two tails. In 16% of pairs of samples (on the left), the female-mean is more than $2\frac{1}{4}$ mm larger than the male-mean; the 16% in the right-hand tail represent male-means that are more than $2\frac{1}{4}$ mm larger than the female. So the probability of a difference larger than $2\frac{1}{4}$ mm is $\frac{32}{100} = 32\%$

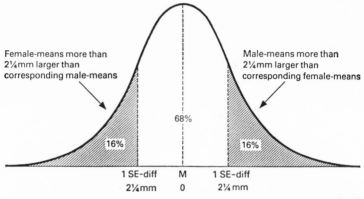

Female-means more than
2¼mm larger than
corresponding male-means

Male-means more than
2¼mm larger than
corresponding female-means

68%

16%   16%

1 SE-diff   M   1 SE-diff
2¼mm   0   2¼mm

Difference between pairs of sample-means (mm)

As we'd expect of a normal distribution, bigger and bigger differences get rapidly less probable, as the curve swoops towards the base-line. From what we know of the normal curve proportions, we can tell that differences as big as $2 \times$ SE-diff will occur in only about 5% of all pairs of samples (because about 95% of the distribution's area lies within Mean $\pm$ 2 SD). So, in the case of our pairs of blood-pressure samples, the probability of a difference in means of $2 \times$ SE-diff (i.e. 2×2.26 mm $= 4.52$ mm) is about $\frac{5}{100}$, or one in twenty. (Remember that, of course, this one pair could have the female mean bigger than the male, or vice versa.)

Similarly, we know that Mean $\pm$ $2\frac{1}{2}$ SD covers about 99% of the normal curve. So, if there actually is no real difference between male and female blood-pressures in general, a difference between means as big as $2\frac{1}{2} \times 2.26$ mm $= 5.65$ mm can be expected in only one pair of samples in a hundred (1% probability). (Again, female bigger than male mean, *or* vice versa.)

What would be the probability of a pair of samples in which the male mean was at least 5.65 mm bigger than the female mean?

*   *   *   *   *   *   *

Since the probability of *either* male exceeding female *or* female exceeding male mean by that amount is $\frac{1}{100}$ (or 1%, or 0.01), then the probability of a particular one of these alternatives is half of that: $\frac{1}{200}$ or $\frac{1}{2}\%$ or 0.005.

We can see these probabilities towards the tails of the distribution curve below:

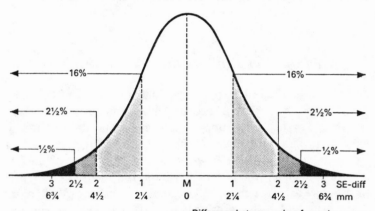

Difference between pairs of sample-means

Perhaps it's time we thought again about our two samples of blood-pressures and our null hypothesis. We said that we'd assume the samples both came from one population – *unless* the difference in their means turned out to be so large that we could not account for it in terms of chance variations among samples from a single population.

Well, the distribution above shows the frequency with which we'd expect the means of pairs of such samples to differ by various amounts (if they came from the same population).

Our two sample-means differed by 10 mm. Would this be a very likely occurrence? Would it lead you to accept the null hypothesis (no real difference), or to reject it?

*   *   *   *   *   *   *

A difference of 10 mm would *not* be very likely. We know that even a difference of 3 × SE-diff would be bigger than 99.7% of all such differences. But here we have one of 4 × SE-diff. The probability of so large a difference could be calculated (using the normal curve tables) as less than six chances in a thousand – so small that we can safely discount the possibility. This means we can reject the null hypothesis. The difference between these samples signifies a real difference between the populations (young men's and young women's blood-pressures). The difference is *significant* – very, very significant, some people might say.

The significance of significance

At this point I'll just remind you that, in statistical thinking, 'significant' does not necessarily imply 'interesting' or 'important.' For instance, suppose we compare a new method of teaching geography with an old method. The mean exam mark of students taught with the new method is, let's say, 66%, while that of a similar sample taught by the old method is 60%. If the samples are large enough, we can establish that this difference is significant. But all we are saying is that we believe the difference to be real, that we believe the new method has put students into a different population from that of the old method – one with a slightly higher mean examination mark. We anticipate that some such difference would be replicated in future comparisons between pairs of samples from these two populations. ('Reliable' would probably be a more appropriate word to use than 'significant' – but it's no doubt too late to change the vocabulary of statisticians now!)

Notice that we are *not* making any value judgements. We are not saying that the difference is worth striving for (or even worth having – geography teachers' reaction to a difference of 6% may well be 'So what!'). We are certainly not saying that the new method (and it might just as easily be a new drug-treatment in medicine or a new production-process in industry) should be adopted in place of the old. This has to be decided on other (non-statistical) grounds, such as: how much does it cost (say, in time and resources) to achieve such a difference? And, would yet other methods achieve bigger differences at less cost?

Which of the following two differences would you say is the more significant, statistically speaking:

(a) The mean sentence-length in novels by author X is 3 words longer than in those of author Y; this difference is calculated as being equivalent to $2\frac{1}{2} \times$ SE-diff.

(b) The survival-rate among patients given a certain operation is 20% higher than that among patients judged to be in need of such an operation but who, for various reasons, have not been able to have it. This difference is calculated as being equivalent to $2 \times$ SE-diff.

* * * * * * *

Statistically speaking, the more significant difference is that in (a). That is to say, it appears that a difference of 3 words per sentence is less likely to happen by chance than is a difference of 20% in survival rate – in the samples compared. Note that it is not the social or human value of the difference we are looking at (nor even its absolute size) – only how big it is in terms of the standard error of such differences (as indicated by variability in the samples to which it relates).

(Remember: the more variable the values in a population, the more variable will be the values observed in samples. And the more variable the samples, the more variable will be their means – and the more variable will be the differences between the means from pairs of such samples. And the *greater the variability* all along the chain, the *bigger the difference between means we'd need to convince us* that our two samples did *not* come from the same population.)

So, a 'significant' difference is one that signifies a real difference in populations. But how big does a difference have to be in order to count as significant? Now this is a bit like asking: 'How big do people have to be to count as "tall"?' There is always something rather arbitrary about drawing a line at a certain value and saying that values on one side of it are significant and on the other side are not significant. Nevertheless, two such 'cut-off points' are commonly used by statisticians. These are the 5% level and the 1% level (often written as the 0.05 level and the 0.01 level).

It may not be obvious at first glance, but a difference is less significant if it is at the 0.05 level than if it is at the 0.01 level. That is, we may accept a difference as significant if it could have occurred by chance (from a single population) in only five pairs of samples in every hundred, or we may insist that it should be so large that it could have occurred in only one pair of samples in a hundred. (We can relate this to the confidence levels we discussed in the last chapter: in the first case we are 95% confident that the difference is a real one; in the second case we are 99% confident.)

So, if we are going to reject the null hypothesis, we will do so more confidently if the difference between our sample-means is significant (a) at the 1% level? or (b) at the 5% level? Which?

* * * * * * *

The bigger the difference, the more confidently we can reject the null hypothesis (that there is no real difference between the means of the populations from which the samples are drawn). To be significant at the 1% level (one chance in 100 of so large a difference), the difference must be bigger than if it were significant only at the 5% level (five chances in 100). So we'd reject the null hypothesis more confidently with a difference significant at the 1% level.

A difference that is significant at the 5% level is often merely called 'significant'; while a difference significant at the 1% level is called 'highly significant'; (and at 0.1% level 'very highly significant'). However, these verbal labels really add nothing to what is shown in the figures – though they may make the researcher feel happier that he has produced something worth while. For instance, suppose two related experiments each produced a difference, one that would have a 4.9% chance of arising merely because of sampling variation, and the other a

5.1% chance. In fact, these two experiments would seem to be in very close agreement about the presence of a real difference; yet in one case it would be labelled 'significant' and in the other 'not significant' (or, at best, 'almost significant'). Perhaps, in evaluating such data, it is better to let the figures speak for themselves.

Let's return to our samples of blood-pressures. How large a difference of means would we have needed for it to be significant at the 5% level? And how large a difference for 1% significance?

* * * * * * *

A difference is significant at the 5% level if it is bigger than $2 \times$ SE-diff. (In this case, $2 \times$ SE-diff $= 4\frac{1}{2}$ mm.) To be significant at the 1% level it must be at least $2\frac{1}{2} \times$ SE-diff (i.e. $5\frac{5}{8}$ mm).

If the difference had been less than $4\frac{1}{2}$ mm, we could not have rejected the null hypothesis. It would simply have been too probable that the difference was due to sampling variations from within one population. In effect, we'd be accepting the null hypothesis – agreeing that a real difference has not been proved. The difference would not be significant.

So, in significance testing, there are two opposite *risks*. First, one may accept a difference as significant when it is not. This is called a TYPE I ERROR. We guard against this by demanding a more stringent level of significance (say, 1% rather than 5%). But, as we increase the significance level (asking for a bigger difference between the sample-means), we increase our risk of making the opposite error.

Can you see what this opposite error would be?

* * * * * * *

If we ask for a bigger difference between sample-means before we'll accept that there is a difference between the populations, then the more likely it is that we'll fail to recognize a difference as being real, even when it is. This is the TYPE II ERROR.

In short: Type I error is to *reject* the null hypothesis when it is *true*; Type II error is to *accept* the null hypothesis when it is *false*. We can picture it like this:

NULL HYPOTHESIS

		True	False
	Reject	ERROR (Type I)	CORRECT
OUR DECISION			
	Accept	CORRECT	ERROR (Type II)

However, since there is no way of knowing whether the null hypothesis is 'really' true or false, we can never tell which of the above boxes our decision puts us in. What we *can* tell is that the more we *reduce* our risk of making a Type I error (by demanding 'more significant' differences), the more we … (a) reduce? or (b) increase? … our risk of making a Type II error. Which?

*   *   *   *   *   *   *

The more we *reduce* our risk of making Type I errors the more we *increase* (b) our risk of Type II errors. The more and more confident we wish to be that we are not going to claim a real difference when there is none, the bigger the difference we'll demand between our samples. But the more we decrease our probability of being wrong in this way, the bigger the differences we'll have decided to *disbelieve* as indicators of real population differences. And, because of the increasing probability that some of these sample-differences will represent real population-differences, the greater our chance of overlooking them and saying two samples are from the same population when in fact they are not.

There is a clear parallel between this dilemma and that involved in the use of evidence in a court of law. If we are prepared to accept rather weak evidence of a person's guilt, then we'll risk punishing a lot of innocent people. Conversely, if we ignore anything but the strongest possible evidence, then a number of guilty people will go unpunished.

What does all this mean in practical terms? If the researcher is very strict in his choice of a significance level, he can be pretty confident of the reality of any differences he *does* recognize; but he may miss many promising possibilities that might have been worth following up with further investigations. On the other hand, if he is lenient in his choice of significance level, he'll be catching all those promising leads, but he may also draw attention to a number of differences on which he (and colleagues influenced by his findings) may subsequently spend a lot of research time and resources, before deciding that the difference was due merely to sampling variation after all. We can imagine how this might apply, say, to some branch of cancer research where nothing approaching a cure has yet been discovered but where the costs of research are extremely high. No doubt the temptation to follow up every faint possibility of a cure would be constantly at odds with the desire to concentrate scarce resources on investigating only the most promising leads (and to avoid raising false hopes among doctors and sufferers). If the costs of a Type I error (rejecting the null hypothesis and mistakenly claiming a real difference) could be expected to be high, researchers would demand a stringent significance level – certainly 1% and maybe even $\frac{1}{10}$% (i.e. 0.01) . . . or even 0.001. It all depends on the circumstances.

Such procedures may seem arbitrary and untidy. But there is no such thing as absolute 100% proof in science (or out of it). All we have is probability and levels of confidence. People with a claim to be scientists are simply more honest about it than others – at least when they are acting scientifically. The risk of error is quantified and stated. We can never know for certain if we have made a Type I or a Type II error, though *further* experiments or inquiries may reduce the uncertainty. But the conventions of science are conservative. They demand significance levels that ensure we are far more likely to *fail* to claim a result that (unknown to us) would have been justified, than to *claim* a result that is illusory.

So the emphasis is on *avoiding* errors of *which* type – I or II?

* * * * * * *

The emphasis is on avoiding Type I errors.

Comparing dispersions

Before we go on to the final section of this chapter, here is a reminder: we've talked mostly about samples differing in mean. But they can also differ in *dispersion* (that is, in standard deviation, or variance). I showed you this diagram at the beginning of the chapter:

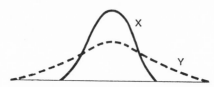

In this diagram, the two samples do not differ much in mean. But they do differ considerably in dispersion. Sample Y is much more spread out, much more variable in its values, much more dispersed, than sample X. There will be a big difference between the *standard deviations* of these samples.

Can a difference in dispersion between two samples be big enough to signify that they come from different populations? Yes, it can. You'll not be surprised to learn that there are ways of testing the significance of a difference in dispersion – but you'll no doubt be pleased to hear that we're not going to look into them! The difference in dispersion pictured above would certainly prove significant; that is, those samples could be assumed to come from two populations differing in variance (though not much in mean).

There are many situations in which we might wish to compare dispersion (e.g. standard deviations) rather than, or as well as, means. For example, anxiety or certain kinds of drug (for example, alcohol) may make some people in a sample better at performing a certain task, while making some worse. Thus, while the mean performance in the sample might not differ much before and after taking the drug, the variability of performance, or dispersion, might well have increased significantly.

As a matter of fact, the significance test we've been discussing so far in this chapter (for the difference in means) does actually depend on an assumption about the dispersion in the samples: it assumes that the pair of samples are sufficiently similar in standard deviation to indicate that they come from populations of *equal* dispersion. The greater the difference in *standard deviation* (or variance) between two samples, the less accurately can we establish the significance of the difference between their *means*.

For example, in which of the three pairs of samples illustrated below would we apply our test for the significance of the difference between means? And in which would we not? Why?

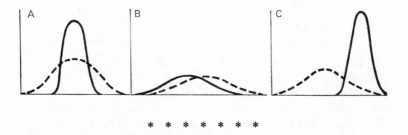

* * * * * * *

The test is effective only on the assumption that dispersion is about equal in the two samples. This is the case only in pair (B) above. So that is the case in which we'd test for the significance of the difference between the means. In case (C) above, the means of the two samples clearly do differ. However, our test would not give an accurate idea of the significance of this difference, because the dispersions are *also* widely different. Nor would it in (A), for the same reason.

When two samples differ in both mean *and* standard deviation, it may be sensible to compare the dispersions first. If the dispersions turn out to be significantly different, then it may be safer to assert that the samples probably do come from different populations – and leave it at that. In other words, it might not be worth carrying out a further test of significance because of the difficulty of separating differences in mean from differences in dispersion. However, we're getting into deep theoretical waters, here – where even professional statisticians may disagree as to what is acceptable practice. So let's move on.

Non-parametric methods

I wonder if it's occurred to you that the statistical techniques we've been talking about seem heavily dependent on the normal curve of distribution? If not, just flip through the pages of this book and notice how that bell-shaped curve appears over and over again – tall and thin, short and fat, and all states in between. It is a fact that most of the 'classical' statistical techniques assume that samples are drawn from normally distributed populations and allow us to estimate the parameters of such populations. Hence, such techniques are called 'parametric.'

But there are many cases where we would definitely be wrong to assume a population is normal in distribution. And there are yet more cases where we may be wrong but can't be sure. For instance, the idea of a normal distribution is inappropriate to category-data. Suppose we are interested in the employment chances of our 'honors' and 'ordinary' graduates. We check up on a hundred of each, one year after graduation, and find that 90 of the honors graduates are employed as are 80 of the ordinary graduates. By a method I'll be describing in the next chapter, it is possible to show that there are less than five chances in a hundred that such would be the proportions if honors and ordinary graduates, in general, are *equally likely* to be employed. That is, we can reject the null hypothesis (at the 0.05 level, anyway). There seems to be a systematic 'non-chance' difference between the two categories of graduate as far as employment is concerned.

But have we made any assumptions about whether employment is normally distributed?

*   *   *   *   *   *   *

Of course we have made no assumptions about the distribution of employment. We have not even measured the amount of employment enjoyed by any individual graduates.

Nevertheless, without any such assumptions, and without calculating means and standard deviations, we are able to test for the significance of a difference in such category-data. This, then, is a NON-PARAMETRIC TEST – it can operate without assumptions about the normality of a distribution.

Another 'category' occasion when only non-parametric techniques can be used is if our data consists of ranks. This is when we know which members of a sample are ranked 1st, 2nd, 3rd, 4th, etc. (e.g. in beauty, intelligence, charm, or whatever). Here the sample members are arranged in order according to 'how much' of the relevant characteristic they are deemed to have. However, these ordered categories may well have been arrived at without any quantity-measurements having been made.

For instance, suppose the students in a college department have arranged the five senior members and four junior members of staff in order of what they regard as their 'teaching ability.' Since no quantity-measurements have been made, there is no way we can average the teaching ability of each group and ask whether there is a significant difference between the mean 'teaching ability' of junior and senior staff. All we know is that the rank order from best to worst teacher is as follows (where S is a senior teacher and J is a junior):

best...								*...worst*
1st	2nd	3rd	4th	5th	6th	7th	8th	9th
S	S	S	J	J	S	S	J	J

However, there is a non-parametric technique (the Mann-Whitney U test) that would allow us to compare this order with the order we might expect if seniors and juniors were equally competent (e.g. SJSJSJSJS) – and, in this case, conclude that the order we have found is not significantly different.

Such non-parametric tests are usually relatively simple to calculate. Consequently, researchers often use them as 'short-cut' (or 'quick-and-dirty') methods, even when they have a full set of measurements and can also assume a normal curve. For instance, they may reduce a set of individual scores on examinations to the numbers of students passing or failing. Alternatively they may forget about the actual scores and simply work on the rank order of the scores.

Naturally, this results in information about the real differences between the members of the samples being lost. For instance, the exam scores of two male students may be 100 and 10, and those of two females may be 90 and 80. So, the male ranks are 1st and 4th, while the female are 2nd and 3rd. The mean rank is $2\frac{1}{2}$ in each case. It seems that women and men did equally well in the exam, if we go by rank-positions alone. But can we still believe in this 'no difference' if we look back at the actual scores made by the men and the women?

* * * * * * *

Well, the two women made a total of $90 + 80 = 170$, while the men's total was $100 + 10 = 110$. Knowing the actual scores has enabled us to see the superiority of the women. But this was hidden when our only information concerned ranks.

So, when we lose detailed information by reducing measured data to a few categories (e.g. pass/fail) or by using rank-positions only, we hide differences. This means we have more difficulty in detecting differences that are significant. Therefore, to put it the other way round, non-parametric tests require differences to be much bigger if they are to be accepted as significant. Thus, although non-parametric techniques are comparatively simple to apply to most statistical inquiries (and must be used in some because parametric techniques are inappropriate), they do increase the risk that we'll accept the null hypothesis when it is false (Type II error).

Perhaps that's enough of a digression on non-parametric methods. They are very important to the practicing statistician, and I'll be mentioning one such method in each of the following two chapters. But statistics is, I believe, best *explained* from the parametric tradition. So the normal distribution will continue to dominate in the remainder of our discussion.

You've probably found this the most taxing chapter in the book so far. Thinking up a theoretical distribution beyond the observed facts; juggling with what seem like the double negatives of the null hypothesis – setting up a result you don't expect just so that you can demonstrate its improbability and thus indirectly vindicate an alternative result you expected all along and might just as well have set out to test directly; expressing the significance of the result in terms of the probability of its being wrong rather than in terms of being correct: all these may make significance testing seem a weird form of mental gymnastics. None the less, it is at the core of statistical method.

In the next chapter I'll give you a chance to improve your 'feel' for significance by taking you, step by step, through a case study in which we'll apply a significance test to an experimental comparison. Meanwhile, if you feel the need to skim through this chapter again before starting on the next, I shan't be in the least surprised. In fact, I'd warmly recommend it!

7. Further matters of significance

In this chapter I want to apply significance testing, step by step, to a new problem. In the process I'll introduce a controversial aspect that I've avoided so far. We'll then go on to deal with more than two samples at once. And we'll end up by considering ways of testing the significance between differences in category- as well as quantity-variables.

Let us suppose we have developed a 'memory drug' which we hope will help students do better in examinations – and, whatever other qualities they test, examinations still do make heavy demands on the memory. We want to know whether students can expect to benefit from taking the memory drug.

We select two random samples, each of 100, from among the students taking a particular examination. Just before they enter the examination room we give the 'experimental' group the 'memory pill.' The other sample of 100 is used as a 'control group.' We give them a 'placebo' – a pill which looks like the real thing but in fact does not contain the drug we are testing. All 200 students are told they have been given the memory drug. (This is to guard against students in the drug group performing differently, not because of the drug, but simply because they have been picked out for special treatment.)

Now, what difference will the drug make to the examination grades of our two samples? (Of course, we'll have to make sure the tests are marked by teachers who are not aware of which students did and did not receive the drug; we wouldn't want to risk their marking being affected by such knowledge. This is called a 'double-blind' in experimental design; neither observed nor observer knows who is who!) Ultimately, we'll want to generalize from the apparent effect on these particular students to its possible effect on other such students in future.

But first, we'll be interested in the mean examination scores of our two samples.

What should be our null hypothesis about these means?

* * * * * * *

The null hypothesis (which we'll hope to find untenable!) is that there is *no significant difference* between the mean score of students who took the drug and of those who did not. (This is tantamount to suggesting that whatever difference appears, it might have arisen by chance, without our drug having had any effect.)

How large a difference are we looking for? Well, let's say there is a good deal of scepticism among our colleagues as to whether such a pill will work. So we feel we can't reject the null hypothesis if the difference turns out to be significant merely at the 5% level. It won't be very convincing if the chance of being wrong is as great as one in twenty.

So we decide we need a difference that is significant at the 1% level if we are to be able to justify further work on the memory drug. This way, we'll be looking for a bigger difference – the chances that we'll claim a difference as significant when it isn't will be down to one in a hundred. Only at the 1% level will we say that the *alternative* hypothesis seems reasonable and worth pursuing.

One- versus two-tailed tests

But what *is* the alternative hypothesis? Here we touch upon a controversial issue that has been rumbling away in the background of statistics for more than a quarter of a century. Shall our alternative hypothesis be merely that there *is* a significant difference, or shall we be more specific and assert that the significant difference lies in a particular *direction*? To be blunt, shall we hypothesize that the 'drug' students will score significantly higher than the 'placebo' students? This, after all, is what we have been hoping and working for. (There are no grounds, after all, for expecting a difference to favor the 'placebo' students.)

What does it matter whether our alternative hypothesis specifies simply 'a significant difference' or 'a significant difference in a specified direction'? Well, think of it in betting terms, in terms of probability:

Which of these two forecasts (a) or (b) is more likely to be correct:

(a) either the drug students will score significantly better than the placebo students, or else the placebo students will score significantly better than the drug students? or

(b) the drug students will score significantly better than the placebo students?

* * * * * * *

The more likely proposition is in (a). After all, it mentions two possible outcomes, and (b) mentions only one of them. (A bettor who says 'horse X or horse Y will be first past the post' is more likely to win than one who mentions horse X only.)

So, if we say in advance that we are looking not simply for a real difference but for a significant difference in favor of the 'drug' students, we would appear to be reducing our chances of success! In fact, we'll make up for this. Let's see how.

If our alternative hypothesis had simply been 'there will be a difference that is significant at the 1% level,' we'd have been looking for a difference big enough to occur in one or other of the two tails of the theoretical distribution of the differences that could be expected even if students were unaffected by being given the drug or a placebo.

As we did with the comparison of blood-pressure means (where we had no reason to suspect the difference would lie in a particular direction), we'd be using a TWO-TAILED TEST of significance.

To reject the null hypothesis (of no significant difference) we'd have required the difference to be so big that there'd be no more than half a chance in a hundred of getting a 'drug' mean this much bigger than a 'placebo' mean, and no more than half a chance in a hundred of getting a 'placebo' mean this much bigger than a 'drug' mean, if the 'two' populations really were one. So the 'CRITICAL' VALUES would have been all those that were *at least* $2\frac{1}{2}$ times the SE-diff (since we know that about 99% of the differences will be less than $2\frac{1}{2}$ SE-diff).

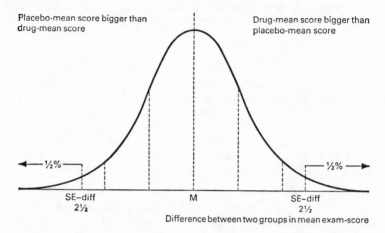

Placebo-mean score bigger than drug-mean score

Drug-mean score bigger than placebo-mean score

½% ◄—   —► ½%

SE–diff   M   SE–diff
$2\frac{1}{2}$ $2\frac{1}{2}$

Difference between two groups in mean exam-score

But that would have been a two-tailed test – appropriate when we have no expectations about the direction of the difference. This time, however, we are posing a more specific alternative hypothesis. We are saying that we will not reject the null hypothesis (and accept the alternative) unless the difference *in favor of the 'drug' students* is so big that it would have no more than 1% chance of being due to sampling variation. In fact, we are about to use a ONE-TAILED TEST. We are interested only in differences where the 'drug' students' mean exceeds that of the 'placebo' students (and *not* vice versa).

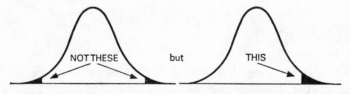

NOT THESE   but   THIS

So, instead of taking the $\frac{1}{2}$% of differences that would lie in the left-hand tail together with the $\frac{1}{2}$% from the right-hand tail, we choose to take *the whole 1% from one tail* – the tail where 'drug' means would exceed 'placebo' means. (In this case, it happens to be the right-hand tail.) We'll reject the null hypothesis only if we get a difference that would be in that right-hand 1%.

How large must a difference in favor of the 'drug' students be if it is to fall in that *biggest 1%* of all such differences? (Check with the curves above.)

(a) Must it be $2\frac{1}{2}$ SE-diff? or
(b) Can it be less than $2\frac{1}{2}$ SE-diff? or
(c) Must it be more than $2\frac{1}{2}$ SE-diff?

*   *   *   *   *   *   *

We'll be looking for a difference in favor of the 'drug' students of (b) less than $2\frac{1}{2}$ SE-diff. In other words, if $2\frac{1}{2}$ SE-diff is about the size of difference that cuts off $\frac{1}{2}\%$ from the right-hand tail, we can see that enlarging that slice to take in an extra $\frac{1}{2}\%$ will push the cut-off point *back* along the base-line, implying a smaller difference.

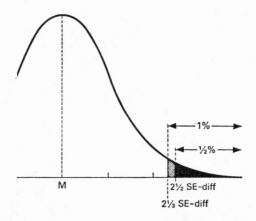

If we look in the tables showing the proportions under the normal curve, we find that the z-value needed to cut off 1% of the distribution in one tail is about $2\frac{1}{3}$. So any difference of $2\frac{1}{3}$ SE-diff or more (in favor of the 'drug' mean) will be in that right-hand tail. To reject the null hypothesis at the 1% level of significance in this one-tailed test, we'll need a difference of at least $2\frac{1}{3}$ SE-diff – in favor of the drug.

So, although we might seem to have restricted our chances of finding a significant difference by sticking our necks out and predicting the *direction* of the difference, we have, in the process, ensured that the difference we need to find is *smaller.** (What we've lost on the swings we've gained on the roundabouts!)

Does this seem to you like sharp practice or sleight of hand? Many people have objected to the use of one-tailed tests, chiefly on the grounds that the experimenter, who knows in which direction the difference should lie, will find it easier to get a significant difference and so reject the null hypothesis. Hence there is a greater risk of Type I error – claiming a difference as significant when it is merely due to sampling variation. And, besides, since such experimenters are ignoring the possibility that the difference may, against all expectations, lie in the *opposite* direction, many important new phenomena may go unrecorded. Perhaps the most telling criticism of all is this: 'If you know so much about the difference that you can even be confident about the direction in which it will lie, why do you need to test for significance at all? A difference so predictable *must* be significant.'

Some objectors also fear that the experimenter who fails to find a significant difference using a two-tailed test may be tempted to switch to a one-tailed test. For although a result of 'no significant difference' might seem a worthwhile scientific outcome to an experiment – helping close off another unpromising line of inquiry – the fact is that such experiments are far less likely to be reported and, if reported, far less likely to be published, than are those in which a significant difference emerged. To some extent, many researchers will feel that lack of significance denotes lack of success.

However much people may criticize them, it seems reasonably certain that researchers will go on using one-tailed as well as two-

---

*By the way, if we'd decided to do a one-tailed test of significance at the 5% level (rather than the 1%), the tables would tell us to look for a difference of $1\frac{2}{3}$ SE-diff or more – rather than 2 SE-diff or more which would have been needed in a two-tailed test.

tailed tests. If we are to be confident about his integrity, when should a researcher state which test he is using (and the level of significance he proposes for rejecting the null hypothesis)?:

(a) before collecting his data? or
(b) after collecting his data? or
(c) at either of the above?

*   *   *   *   *   *   *

The researcher should say whether he is using a one-tailed or two-tailed test (and the level of significance at which he will reject the null hypothesis) *before* collecting the data. (Bookmakers would have little sympathy for someone who tried to place his bet only once the race had been run!) If, for example, the researcher used a two-tailed test to discover whether there was any reliable difference at all, he should not switch to a one-tail comparison in order to report a higher level of significance. What he should do, in such circumstances, is to collect *more* data with new samples, and apply a one-tailed test to that.

Unfortunately, it is not always possible to tell whether a researcher's alternative hypothesis was formulated prior to the experiment, or was somehow tailored to fit the results themselves. In interpreting other people's reports we must be wary about accepting their claims of 'significance' at face value. We should consider first the background of the experiment (for example, the results of similar experiments elsewhere, and the theoretical reasonableness of the result). Then we can ask: does the probability of the recorded difference arising purely out of sampling variation seem low enough to justify expecting it to occur reliably in future comparisons?

So let's find out how our examination candidates performed. What was to be the *null hypothesis* in this experiment?

*   *   *   *   *   *   *

The null hypothesis, you'll remember, was that the mean score of the 'drug' students would not be significantly different from that of the 'placebo' students. Our alternative hypothesis (formulated in advance of comparing the results, you'll notice) was that the 'drug' mean would be higher.

The next step in significance testing is to specify the theoretical distribution of differences between means we'd expect from a population *if* the null hypothesis *were* really true. That is to say, we'd expect a normal distribution with a mean of zero (no difference) and a standard deviation estimated by combining the standard errors of the two means we'll get from our samples. (In other words, the greater the dispersion in the samples. the greater will be their standard deviations; the greater their standard deviations, the greater will be our estimate of the standard errors of their means; and the greater will be our estimate of the standard deviation, or standard error, of differences between means from such samples.)

We now specify the level of significance at which we'll reject the null hypothesis. We decided, you remember, on a significance level of 1%. This implies we want a difference between means (in favor of the 'drug' mean) so large that the probability of its occurring by chance from the theoretical 'no difference' population mentioned above would be 1% or less. Thus we can indicate a 'CRITICAL REGION' in the theoretical distribution, where the alternative hypothesis will seem more acceptable to us than the null hypothesis. In this case, the critical region will fall in the tail of the distribution where 'drug' means exceed 'placebo' means by at least $2\frac{1}{3}$ SE-diff.

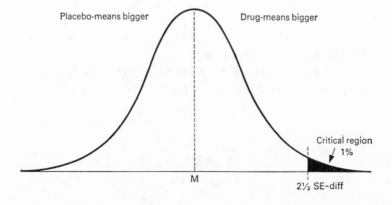

Now we can look at the actual examination scores. We calculate the means and standard deviations for the two samples of
students to be as follows:

	mean score	standard deviation
'Drug' students	62.8	10 marks
'Placebo' students	60	9 marks

How big is the difference in mean scores? Which of the two
groups does it favor?

* * * * * * *

The difference in mean scores is 2.8 marks and it does favour the
drug students. But is it a big enough difference to fall in what
we've decided should be the critical region? To answer this, we'll
need to know the standard error of the difference between the
means. This we'll get by combining their separate standard
errors.

So, the standard errors of the two means are as follows
(remember, there are 100 students in each sample):

$$\text{SE-mean (drug)} \quad = \frac{10}{\sqrt{100}} = \frac{10}{10} = 1 \text{ mark}$$

$$\text{SE-mean (placebo)} = \frac{9}{\sqrt{100}} = \frac{9}{10} = 0.9 \text{ marks}$$

(Just to remind you what SE-means are about, the above figures
imply that about 68% of any large group of similar students would
score between 62.8 ± 1 marks if given the drug, or between
60 ± 0.9 marks if given the placebo.)

The standard error of the *difference* between the means is a
combination of their separate standard errors:

$$\text{SE-diff} = \sqrt{1^2 + 0.9^2}$$
$$= \sqrt{1 + 0.81}$$
$$= \sqrt{1.81}$$
$$= \quad 1.3 \text{ marks}$$

We can now indicate the theoretical distribution of differences that would exist *if* the null hypothesis *were* true and both sample-means came from the same population.

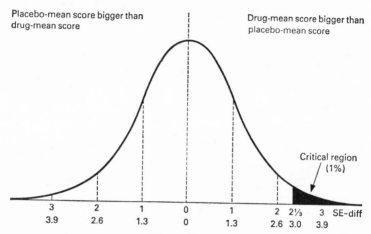

Difference between two groups in mean exam-scores

Since SE-diff = 1.3 marks, we can work out that 2 SE-diff = 2.6 marks, 3 SE-diff = 3.9 marks, and so on. But the value of SE-diff we are particularly interested in is $2\frac{1}{3}$ SE-diff, because this is the value that cuts off the critical region. Our difference in means has to be larger than this if it is to fall in that region and thus be significant at the 1% level.

Well, $2\frac{1}{3}$ SE-diff = 2.3 × 1.3 = 3 marks. *Is* the difference between the mean scores of 'drug' students and 'placebo' students (that is, 2.8 marks) significant at the 1% level?

* * * * * * *

Since the difference between the mean scores of 'drug' students and 'placebo' students is 2.8 marks, we have to conclude that it is *not* significant at the 1% level – not quite! (To get in the 1% critical region, we'd have needed a difference of at least 3 marks.)

This difference of 2.8 marks would, of course, have been significant at the 5% level. As you'll remember, in a one-tailed test, the 5% critical region would have included all differences of at least $1\frac{2}{3}$ SE-diff. Since, in this case, $1\frac{2}{3}$ SE-diff $= 1\frac{2}{3} \times 1.3 = 2$ marks, our difference of 2.8 marks would have been well into the 5% critical region.

This is just one of those cases where we might be tempted to retreat into meaningless verbal labels – saying the difference is not 'highly significant' but it is 'significant'! Common sense, however, suggests that we look at the figures themselves before making a decision. Admittedly the difference was not quite big enough to have a less than 1% probability of occurring by chance. So we can't be 99% certain that the two means come from different populations, and thus we can't reject the null hypothesis. However, the difference did come *very close* to the critical region.

Suppose the same difference had been found between two much larger samples – say of 150 students each. Would its significance have been the same?

* * * * * * *

No, if the same difference had been found in a pair of much larger samples, its significance would have been greater. With samples of 150 students each (which would reduce the standard error to about 1.1 marks), the difference would have been significant at the 1% level.

So, although there is insufficient evidence to be sure that 'drug' students can be relied on to do better than 'placebo' students, we may take reassurance from the fact that the weight of the evidence does nevertheless favor such a possibility. Hence, if the drug was not too expensive to prepare, and appropriate examination candidates not too difficult to get hold of, we might, after all, try the experiment again with larger samples. On the other hand, we might well decide that, even if it turned out to be real, a difference of just two or three marks wouldn't be big enough to justify further work. What we make of a difference (given its probability of having arisen by chance) must depend not only on which side of the 'magic' 5% or 1% it lies, but on many other, non-statistical factors also.

z-tests and t-tests

By the way, the significance test we've spoken most of so far is sometimes called a z-TEST. This is because we use the standard deviation as a unit (z-unit) for measuring the point where the critical region begins (e.g. 2 SD = 5% or $2\frac{1}{2}$ SD = 1%, for two-tailed tests) *and* relate it to the proportions of the normal curve.

However, this is accurate only with large samples. When samples contain less than 30 members, the sample standard deviation can no longer be relied on as an estimate of the standard deviation in the population. (You'll remember, from page 92, that the calculation of the standard error of the mean, on which the whole structure depends, is based on such a reliance.) Indeed, William Gossett, who published under the pen-name of 'Student,' noted that, in small samples, sample-SD *under*-estimates population-SD more than half the time.

Consequently, with samples of less than 30, Student's *t-test* is used instead. I won't trouble you with the details of this. But you'll come across the term, so you'd better have a general idea. In fact, the *t*-test uses the standard error of the differences between means (calculated exactly as in a *z*-test, but re-named the '*t*-value'). But, for the necessary proportions either side of a given value, we refer not to the normal distribution but to the *t-distribution*.

The *t*-distribution is similar to the normal distribution in being symmetrical about a mean of zero, and bell-shaped. But it is flatter (more dispersed) and its dispersion *varies* according to the size of the sample. The diagram on the next page shows the normal distribution and the *t*-distributions for samples of 6 and of 20. As you can see, the bigger the sample, the more nearly does the *t*-distribution approach the normal distribution.

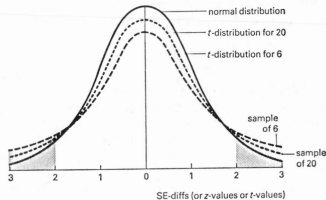

SE-diffs (or z-values or t-values)

Look at the area beyond 2 SE-diff in the graph above. Compare the proportion of the differences (the area under the curves) in each of the three distributions. Clearly, the proportion of *extreme* differences between two samples is assumed to get greater and greater as the sample size gets smaller. That is, small samples are more likely to give significant-looking results.

So, to establish significance at a given level (e.g. the 5% level), the *t*-test for small samples must demand what kind of difference: (a) a smaller difference or (b) a larger difference?

* * * * * * *

The smaller the samples, the *larger* the difference demanded by the *t*-test; otherwise, since big differences occur more frequently as samples get smaller, the probability of getting a significant result would increase as the sample size decreases.

With the *t*-test, then, a big difference is needed to establish significance if the samples are small. As the sample size increases, smaller and smaller differences become significant. For instance, to be significant at the 5% level, the difference between samples of 4 would need to exceed 3.2 × SE-diff. But with samples of 12, a difference of 2.2 × SE-diff would be significant at this level. Then, with samples of around 100, the difference need only reach $2 \times$ SE-diff. With such large samples, the t-distribution becomes identical with the normal distribution.

Comparing several means

The z-test (or t-test) is used when we wish to check whether *two* samples are likely to have come from the same or from different populations. But very often we are interested in the differences between three or more samples. For instance, though we compared the mean scores of 'drug' students and 'placebo' students, perhaps we should have compared their mean scores with that of a third sample group – a random sample from among those examination candidates who were given neither drug nor placebo. (After all, the 'placebo' students may have performed better than untreated students simply because they *thought* they had been given the drug.) Similarly, we may wish to know whether there are any significant differences in learning ability between four breeds of mice . . . five different methods of teaching quadratic equations . . . six different conditions for the germination of seeds . . . and so on.

At first sight, you may not be aware that there's a problem here. Why not simply compare each possible pair of samples, using a z-test or t-test? Thus, for example, the drug sample could be compared with the placebo sample and with the no-treatment sample, and the placebo sample could be compared with the no-treatment sample: three significance tests in all. Similarly, if we were looking for significant differences between samples A, B, C and D, we could compare them like this: A–B; A–C; A–D; B–C; B–D; C–D; six tests in all. Unfortunately, as we increase the number of samples to be compared, the number of inter-sample comparisons increases at a much faster rate. Thus, six samples would involve 15 tests, eight would involve 28, ten would involve 45, and so on.

Now the sheer labor of doing all these tests is one disincentive. How frustrating to do, say, 45 inter-sample comparisons before discovering that there is no significant difference anywhere among them – especially since there is one test that enables you to check first whether there is such a difference at all before you start trying to track it down.

But there is an even more important reason for doing one test rather than several. Remember that, if we are looking for a difference that is significant at, say, the 5% level, we risk that in the long run we'd be *mistakenly* rejecting the null hypothesis five times in a hundred. That is to say, because of chance variations in samples, one out of every twenty comparisons is likely to reach the 5% level of significance by *chance alone*.

So, as we *increase* the number of z-tests (or t-tests) we do on a body of data, how does it affect our chances of making a false claim?

* * * * * * *

The greater the number of separate tests we make, the greater is the likelihood that we'll claim some differences as real when actually they are due to chance.

All right, so multiple tests can't be justified. Another solution that might occur to you is to take the pair of samples with the biggest difference in means, and apply the significance test to them. (And perhaps, if that difference proves significant, you next take the pair with the second biggest difference, and keep on until you reach a difference that is no longer significant.) However, to do this would be to load the odds unfairly in favor of finding a significant difference. Can you see why?

* * * * * * *

Well, the significance test applies only to the difference between pairs of samples chosen *randomly* from the population. Clearly, if we choose a pair of samples *because* they are very different, then we are not choosing at random.

After all, it should be apparent that the chance that the biggest of several differences will turn out to be significant is much greater than that of a single difference whose size you don't yet know. (Similarly, if you predict merely that the *winner* of the next Olympics pole-vault will break the world record, you are on to a safer bet than if you have to forecast the *name* of the record-breaker.) The problem resembles that with one-tailed versus two-tailed tests. You would be justified in testing individual differences only if you'd predicted the ones likely to be significant *in advance*.

Fortunately, there is a single test whereby several samples can be compared at once. It is called the F-TEST – after the British statistician, R. A. Fisher, who developed the process called ANALYSIS OF VARIANCE, on which it rests. Analysis of variance could be applied to comparing two groups. In such a case, the F-test would give the same result as a *z*-test or *t*-test. It is more commonly used, however, when three or more samples are to be compared. It answers the question: 'Are there one or more significant differences *anywhere* among these samples?' If the answer is No, there will be no point in examining the data any further.

The computations involved in analysis of variance can be quite complex.   Nevertheless, since it is such an important technique, I really must try to convey some idea of its underlying principles.

The basic situation in which we use the technique is this: we have a number of observed values, divided among three or more groups. What we want to know is whether these observed values might all belong to the same population, regardless of group, or whether the observations in at least one of the groups seem to come from a different population. And the way we'll get at an answer is by comparing the variability of values *within* groups with the variability of values *between* groups.

I'd better just mention, in this connection, that the F-test makes a similar assumption to the *t*-test (or *z*-test) which is: the samples are *all* supposed to have similar dispersions.

In pictures, then: which of the two situations below would analysis of variance (the F-test) apply to?

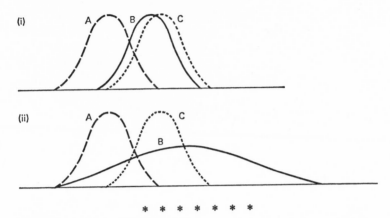

* * * * * * *

Analysis of variance would apply to situation (i), in which the dispersion is approximately the same in *all* samples. But not to situation (ii), in which one of the samples is quite different in dispersion.

Now let's illustrate analysis of variance with an example – an extreme one to make the point. We have given a test to three groups of students. But first, we gave our memory-drug to one group (A), a placebo to a second group (B), and did nothing at all to a third group (C). Their test-results (out of 30) show the following (somewhat striking!) pattern:

	Group A	Group B	Group C
	21	11	1
	23	13	3
Students' scores	25	15	5
	27	17	7
	29	19	9

Compare the variability of scores *within* groups with that *between* groups. Would you, or would you not, feel that one or more of these groups was significantly different from the others?

* * * * * * *

The variability of scores within groups is small. It happens to be the same in each case, with scores differing by no more than four points either way from the group-means of 25, 15 and 5. (In fact, the standard deviation of each group is 3.16*, suggesting a similar standard deviation in the population from which they came.) But clearly the scores differ much more from group to group than they do within groups. There is no overlap between the groups. Group A scores are all in the twenties, Group B in tens, and Group C in units. (In fact, the standard deviation (standard error) of the three sample means (25, 15 and 5) is 10.†) If these means were all to come from the same population of sample-means, this standard error suggests that the population would need to have a standard deviation of about 22 – far bigger than the population-SD suggested by the variability within the groups. So we can feel pretty sure that all three groups are significantly different from one another – they come from different populations. We'd expect similar group differences in similar circumstances. We can predict that, on average, students given the memory-drug will do better than students given a placebo, and both will do better than students who are left alone.

So the analysis of variance depends on comparing the variability of values within groups with the variability between groups. We ought to be asking whether the random variability between individuals seen in the samples was *sufficient* to account for the variability between samples. In a case like that above, we wouldn't need to apply analysis of variance, because the answer leaps out at us. Let's look briefly at another example, still very simple, but less blatant, and follow through the logic of analysis of variance.

---

*With samples as small as this, to divide the sum of the squared deviations from the mean by the number in the sample (in this case 5) would be to *under*-estimate the population-SD. It is customary to divide by one less than the number in the sample, so I have divided by 4.

†Again, I have divided the sum of the squared deviations from the overall mean by one less than in the sample of three means, i.e. 2.

Suppose our three groups of test students had produced the following (rather more ambiguous) scores:

	Group A	Group B	Group C
Students' scores	25	24	20
	27	24	22
	28	26	24
	30	28	24
	30	28	25

It is not so easy here to compare variability within and between the groups. We cannot tell, just at a glance, how different the groups are from one another. If you examine the scores closely, however, you should be able to see which group has the largest mean and which the smallest. Which is which?

* * * * * * *

The three means are: A–28; B–26; C–23.

Are these means sufficiently different (taking account of the variability in all the data) to suggest that they come from more than one population? Analysis of variance will tell us.

We'll begin with the null hypothesis that all three samples are from the same population. The alternative hypothesis is that they are not all from the same population. That is, it could be that one of the samples comes from a population whose mean is different from that of the population the other two samples come from. Or it could be that all three samples come from populations with different means.

Anyway, with the data we have available, we can make *two different estimates* of variability in the common population we are assuming under the null hypothesis. According to how closely these two estimates agree, we accept or reject the null hypothesis. For computational reasons, which I won't bother you with, this variability is expressed not by a standard deviation but by the square of the standard deviation. This, you may remember (from page 54) is called the *variance*. So $V = SD^2$.

We get our first estimate of the variability in the population by pooling the variability from *within* the three samples. Calculations would show that the average variance within the samples is 4.17.

Our second estimate of the variability in the population comes from the three sample-means. The variance among them is only 6.33; but, if they are all supposed to have come from one population, its variance would have to be much larger. In fact, using the relationship between SE-mean (which we can calculate directly from the sample-means), sample-SD, and sample-size – which we last looked at on page 92 – the population variance can be estimated as 31.65.

So we have two estimates of variance in the common population we assumed under the null hypothesis. To assess the null hypothesis, we'll need to compare the two estimates.

Which of the two estimates of variance would you expect to be larger if we are to reject the null hypothesis (and say that the samples come from more than one population)? Would it be (a) the within-groups estimate? or (b) the between-groups estimate?

* * * * * * *

If the null hypothesis is false (and there is a real difference in population-means) we can expect the between-groups estimate of variance (b) to be larger than the within-groups estimate. That is, it will be suggesting a variability greater than the mere sampling variation suggested by the within-groups variance.

In this case, the between-groups estimate of population variance (31.65) is clearly much larger than the within-groups one (4.17). But, is it *large enough* for us to be confident that it is attributable not simply to random, sampling variation but also to the fact that at least one of the samples contains values that are, on the whole, significantly greater or smaller than one of the other samples? (In short, we're asking whether the difference between these samples is a *reliable* one that we could expect to see repeated in similar tests.)

Next, we compare the two estimates by calculating the ratio between them – dividing the between-groups estimate by the within-groups estimate – and observing whether it is greater than 1. We then check this variance-ratio (or F-RATIO) against the

F-DISTRIBUTION. This is a 'family' of curves, like the t-distributions, whose shape differs according to the samples. The F-curves, however, differ not only according to the *size* of the samples but also according to *how many* samples are being compared. In general, the smaller the samples (and/or the fewer of them), the bigger must the F-ratio be in order to attain significance.

Of course, there are tables of the F-distribution, giving the critical ratios for all sample-sizes and numbers of samples. If we looked in these tables, we'd see that, with three samples of five students, we'd need an F-ratio of at least 3.89 in order to reach significance at the 5% level, and an F-ratio of at least 6.93 to reach the 1% level of significance.

Our two estimates of population variance give an F-ratio of

$$\frac{31.65}{4.17} = 7.59$$

which clearly exceeds 6.93 and indicates that the two estimates are significantly different at the 1% level. How can we interpret this difference:

(a) all three samples are from the same population? or
(b) the samples are not all from the same population? or
(c) the three samples are all from different populations?

* * * * * * *

It is not possible to say whether the sample-means are all significantly different or whether two of them are the same. All we can be sure of is (b), the samples are not all from the same population. Thus we reject the null hypothesis.

So, as we've seen in the above example, analysis of variance is a technique used to analyse the overall variance in a set of samples. The variance is divided into two components: the within-samples variance and the between-samples variance. If the between-samples estimate of variance is much bigger than that from within samples, we conclude that ordinary sampling variation is not enough to account for the difference in means. At least one of the samples comes from a population with a different mean.

The analysis of variance can be applied to much more complex comparisons than the example given above. That was a ONE-WAY ANALYSIS OF VARIANCE. But we could have two-way, three-way, four-way, etc., analysis of variance. As an example of a TWO-WAY ANALYSIS OF VARIANCE, we might re-run our test (but with a larger sample), noting the scores not only according to whether students were given the drug, the placebo, or nothing, but also according to whether they were, say, male or female.

This would give us six groups of students. Each group might have a different mean score on the test, something like this perhaps:

	Drug	Placebo	Nothing
Male	25	26	22
Female	28	22	25

Looking at the variability of the scores within the groups as well as between them, we could then ask whether the means are significantly different for men and women as well as whether they are significantly different for the three conditions originally compared. We might also have classified our test students as, say, introvert or extrovert, and thus done a three-way analysis of variance. (We'd then have needed $3 \times 2 \times 2 = 12$ groups to ensure we had reasonably large samples. It should be clear that we have to be very careful about how we design our experiment if there is any intention of using analysis of variance on the resulting data.)

Finally, on analysis of variance, we might notice that once we get on to two-way, three-way, etc., we can look for INTERACTION EFFECTS between conditions. For instance, in the table above we see that how students do on the test depends not only on how they are treated beforehand, but also on which sex they are. If we brought in the personality dimension, we might discover that, say, introvert males responded to the memory-drug more like extrovert females than like éxtrovert males, and so on. By analysing the overall variance of the students' scores into its separate components, we could test any such hypotheses that seemed worth following up.

Comparing proportions

To end this chapter, I must just mention a very important kind of significance test that is used when our investigation concerns category-variables rather than quantity-variables (that is, when we are concerned not with a measurement of quantity for each member of the sample but with counting how many members fall into each of a number of descriptive categories). This is one of the 'non-parametric' tests I mentioned back on page 125.

The simplest comparisons would concern characteristics with just two categories: e.g. male/female; yes/no; pass/fail; plain/fancy, and so on. For instance, one characteristic might concern the sex of applicants to college, and the second might concern whether or not they have had previous work-experience (as defined in a questionnaire). We might then test whether there is a significant difference between the proportions of males and females who have had prior work-experience.

In fact we could test the significance of such a difference by using a statistic mentioned on page 99: the standard error of a proportion. There is a test for the significance of a difference between two proportions that closely follows the method we discussed for testing the difference between two means. However, I want to introduce here an alternative method, which can be used with attributes that have *more than two* categories. It is, in fact, one of the most widely used tests in social statistics. It is called the χ^2 test. (That, by the way, is not a large X but the Greek letter, chi, pronounced 'Kye' as in 'sky'.) You'll see it written also as the CHI-SQUARE TEST.

How would we use this test to compare the work-experience of male and female applicants? Consider the data we obtain from a random sample of applicants, 100 of each sex:

Obtained frequencies

		Work Experience		
		yes	no	TOTAL
Sex of	male	70	30	100
Applicant	female	50	50	100
	TOTAL	120	80	200

Can you see any signs of a correspondence between an applicant's sex and whether or not he/she has had prior work-experience?

* * * * * * *

The table does seem to show a correspondence. 70% of the men have had work-experience compared with 50% of the women.

However, what we have to decide is whether this represents a real, reliable difference between the proportions in the populations of male and female applicants. How far can we be sure that it's not just a chance, sampling variation that could be wiped out, or even reversed, if we took another sample of 200 applicants?

We'll start, as always, by stating our null hypothesis: men and women applicants are equally likely to have had work-experience. On the basis of this null hypothesis we draw up a table of EXPECTED FREQUENCIES. That is, if men and women are really equally likely to have work-experience, how many of the 200 students would we expect to fall into each of the four cells of the middle table? (If we are to assert that there *is* a real difference between men and women on this attribute, then we must demonstrate that the gap between the *expected* frequencies and those we actually *obtained* is just too big to have arisen out of sampling variation alone.)

So what are the expected frequencies? How would the 200 students be expected to parcel themselves out if there really were no sex differences? Surely, we would expect to see the percentage of experienced men equal to the percentage of experienced women. And both should be equal to the percentage of experienced applicants in the sample as a whole, i.e. 60%.

Expected frequencies

| | | Work Experience | | |
		yes	no	TOTAL
Sex of	male			100
Applicant	female			100
	TOTAL	120	80	200

We saw from our sample that 60% of applicants overall (120 out of 200) had work-experience. So how would you fill in the four middle cells of the table to show what we'd expect, *if* this proportion were true also for the men and women separately?

*   *   *   *   *   *   *

We'd expect to see 60 men in every 100 with work experience and 40 without; and the same for 100 women.

Expected frequencies

| | | Work Experience | | |
		yes	no	TOTAL
Sex of	male	60	40	100
Applicant	female	60	40	100
	TOTAL	120	80	200

Now we'll put our expected (E) and obtained (O) frequencies together and see how much they differ. The bigger the difference, the less likely it is to have arisen within a single population.

| | | Work Experience | |
		yes	no
Sex of Applicant	male	O: 70 E: 60	O: 30 E: 40
	female	O: 50 E: 60	O: 50 E: 40

As you can check, the difference between the obtained and the expected frequency in each cell is +10 or −10. Is this big enough to be significant? The way we decide is by calculating the chi-square statistic. The size of χ^2 depends on the sizes of the differences and (loosely speaking) on how many differences are involved.

For the data above, χ^2 turns out to be equal to 8.33. If the differences between observed and expected frequencies had been greater, then chi-square's value would have been more than 8.33. (Similarly if there had been more categories involved.) On the other hand, smaller differences would have made χ^2 smaller than 8.33 But, again, is 8.33 big enough to be statistically significant?

You won't be surprised, by this time, to hear that we'd look for the answer in a table showing the distribution of chi-square. The chi-square distribution, like that of t and of F, is a *family* of curves; their shapes vary (loosely speaking) according to the number of differences involved. When we look at the table for distribution of chi-square involving four differences, we find that a χ^2 of 3.84 is exceeded in only 5% of occasions, and a χ^2 of 6.33 in only 1%.

So, is the χ^2 we have found – 8.33 – significant? And, if so, at what level? And what does this result tell us about the likely work-experience of men and women applicants generally?

* * * * * * *

The value of chi-square we obtained, 8.33, clearly exceeds not only 3.84 but also 6.63. We can say, therefore, that it is certainly significant at the 1% level. A value this big should occur less than once in a hundred samples if there really were no difference in work-experience between men and women applicants. So we are prepared to reject the null hypothesis and assert (with 99% confidence!) that the difference is real – we can rely on its appearing again in similar samples.

Enough of significance testing. You will no doubt have found the last two chapters quite tough going. The next, and final, chapter should seem considerably easier. In our final chapter we'll look at the important statistical concepts of correlation and regression. But you'll find this gives us an opportunity to review many of the ideas we've touched on earlier.

8. Analysing relationships

In previous chapters we have considered how to describe samples in terms of some category- or quantity-variable. We have also considered how to generalize (with proper caution) from a sample to a population. Furthermore, we have compared samples with a view to inferring a difference between populations. In particular, we have seen how to show whether two or more samples that are different in one variable are *also* significantly different in some other variable. For instance, we know how chi-square is used – as we did to establish whether students who have prior work-experience are more likely to pass the first year of the course than students who have none. And we understand the use of a *t*-test (or a *z*-test) – for example to establish whether the students whose travelling-time to college is above average have a significantly higher absentee-rate than students whose travelling time is less than average.

However, such comparisons merely establish a connection. They tell us something about students in general, but they do not allow us to make very precise predictions about individuals. This is because so much information has been lost by reducing the data to dichotomies (above/below average; pass/fail, etc.), and by representing a spread of values by one 'typical' value, the mean. In fact, most of the procedures we've discussed so far are appropriate chiefly to exploratory research: identifying connected variables.

But we often want to go beyond this. We want to establish the nature of the relationship between two or more variables. For some reason or other, we may also want to *predict*, say, the following: given that a certain student has so many months of work-experience, what percentage is she likely to get in her first-year exam? Alternatively, given that another student lives so many minutes away from the college, how often is he likely to be late during the term? Or, again, given that a student has such-and-such a pulse-rate, what is his blood-pressure likely to be? And, in all such situations, the accuracy of our predictions would depend on the *strength* of the relationship.

This takes us into the study of CORRELATION and REGRESSION. Correlation concerns the strength of the relationship between the values of two variables. Regression analysis determines the nature of that relationship and enables us to make predictions from it.

Paired values

In studying correlation, we'll be looking at samples where each member has provided us with a value on *two* (or more) different variables. For instance, we might test the intelligence and the manual dexterity of thirty students and so get thirty pairs of values. We might compare crime-rate and unemployment-rate for each of the country's twenty largest cities. We might compare the distance and the brightness of a hundred stars. In each case, we can look to see how far large values on one variable are associated with large values on the other, or whether large values on one are linked with small values on the other.

So, from each member of the sample, we get two different measurements. Taking a very clear-cut example to start with, suppose our sample consists of a number of circles of differing radii. For each circle we know the radius and the circumference. Here are the figures:

Circle	A	B	C	D	E
radius (cm)	1	3	5	8	10
circumference (cm)	6.28	18.85	31.42	50.27	62.84

How would you describe the relationship between these two variables – what happens to the values of one variable as those of the other increase or decrease?

* * * * * * *

Clearly, the circumference increases in length when the radius increases. Big values on one variable are associated with big values on the other, and small with small. Variables that vary together are said to be correlated – related one to the other.

We can illustrate the exactness of this particular relationship by plotting the values on a new kind of dot-diagram. Each dot on the diagram below represents a sample member, for which we have *two* measurements.

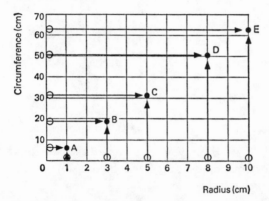

Radius (cm)

(You can, if you like, see this as a *combination* of two dot-diagrams of the type I used in Chapter 3 to describe distributions. To show you what I mean, I've 'dotted' the distribution of radius sizes along the horizontal axis, and that of the circumference up the vertical axis. Then I've pulled these up and across, respectively, to form the common distribution of *paired* values.)

Once we are satisfied that some relationship does exist, we can establish the precise nature of that relationship and use it to predict values of one variable that would correspond to any given values of the other. In the case of our circles, we know that radius and circumference are related by a formula ($C = 2\pi r = 2 \times 3.142 \times r$). So we can 'predict' exactly the circumference of any circle, given its radius, e.g. a circle of 20 cm radius will be $2 \times 3.142 \times 20 = 125.68$ cm in circumference.

The exactness of the relationship can be seen in the diagram above. The plotted points lie along an imaginary straight line. If we were actually to draw a line through them, we could use it for making exact predictions without even using the formula.

Most correlations are not so clear cut, however; hence our

predictions can be nothing like so accurate. Suppose we take a sample of ten students (A–J) and compare their marks on two different tests, one theoretical, one practical:

Student	Theory	Practical
A	59	70
B	63	69
C	64	76
D	70	79
E	74	76
F	78	80
G	79	86
H	82	77
I	86	84
J	92	90

What relationship can you see among these pairs of scores? Are high theory marks associated with high or low practical marks? Always, or just usually?

＊ ＊ ＊ ＊ ＊ ＊ ＊

With a few exceptions, students with high theory scores also did well on the practical. However, it was not too easy to see this pattern in the raw data. Let's try to bring out the relationship with a dot-diagram:

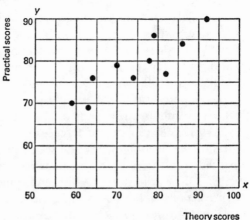

(Note that it is customary to label the variable on the horizontal axis *x*, and that on the vertical axis *y*.) Check that you can relate this diagram to the table above. Each of the ten dots represents the pair of scores made by one of the ten students. For instance, which student is represented by the dot in the top right-hand corner of the diagram?

* * * * * * *

The dot in the top right-hand corner stands for Student J – who made scores of 92 (theory) and 90 (practical). Just to be sure, I have now put the name of its 'owner' alongside each dot on the version of the diagram below. I have also drawn a straight line through the dots; that I'll explain in a moment.

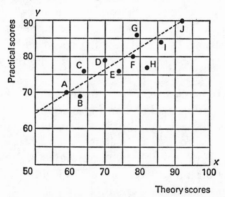

Theory scores

This diagram shows quite clearly that there *is* a relationship between how students score on the two tests. Higher practical scores tend to go with higher theory scores, and lower with lower. But there are exceptions. The relationship (the correlation) is much less strong than that between the radius and circumference of circles.

Because the correlation is weaker, the dots do *not* lie along a straight line. Admittedly, as I've done in the diagram above, we can 'imagine' a straight line running through (among) the dots – from bottom-left to top-right. But the dots would be *scattered* away from any such line, with some on either side of it.

In fact, it is usual for this to happen in a dot-diagram illustrating a correlation. The dots are usually scattered on either side of an imaginary straight line. Hence, such a dot-diagram is called a SCATTER DIAGRAM.

Three kinds of correlation

The scatter diagram of scores above shows an example of what is called POSITIVE CORRELATION. There, changes in one variable are accompanied by changes in the other variable and in the *same* direction; that is, larger values on one variable tend to go with larger values on the other.

But in many, equally strong relationships, the two variables change in *opposite* directions. Larger values on one will tend to go with *smaller* values on the other. This is called NEGATIVE CORRELATION. There might well, for example, be a negative correlation between age and running speed within a group of college lecturers, i.e. the older, the slower; and the younger, the faster.

If there were no clear tendency for the values on one variable to move in a particular direction (up or down) with changes in the other variable, we might say we had something approaching a ZERO CORRELATION. (I say 'approaching' because it is very difficult to find two variables with absolutely no relationship at all.) The relationship between age and weight of adult students might well approach zero correlation.

Here are three scatter diagrams. They illustrate the three kinds of correlations mentioned above. Which is which?

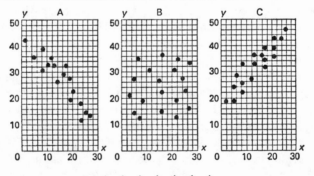

*   *   *   *   *   *   *

The correlations illustrated in the scatter diagrams above are negative (A), approximately zero (B), and positive (C).

Now let us determine the correlation between each of the following pairs of variables. Mark each pair to show whether you'd expect each correlation to be positive (+), negative (−), or towards zero (0).

 (i) Rainfall and attendance at football games.
 (ii) The age of a car and its value.
 (iii) Length of education and annual earnings.
 (iv) Sales of television sets and cinema attendances over a period of time.
 (v) Ability to see in the dark and quantity of carrots eaten.
 (vi) Level of unemployment and installment plan sales over a period of time.
 (vii) Miles driven and amount of fuel consumed.
 (viii) Amount of smoking and incidence of lung cancer.
 (ix) The scores obtained on each of two dice thrown together.

* * * * * * *

I would expect the following correlations:
(i) −. (ii) −. (iii) +. (iv) −. (v) 0. (vi) −. (vii) +. (viii) +. (ix) 0.

The strength of a correlation

As I have already mentioned, correlations vary not only in direction (+ or −) but also in *strength*. We saw the strongest possible relationship – a *perfect* relationship – between the radius and the circumference of circles. Here the plotted points on the 'scatter' diagram lay on a *straight line*.

Similarly, of course, we could have a perfect negative relationship: say, between the size of a withdrawal you make from your bank account and the amount that would be left to your credit. There, too, the dots on the 'scatter' diagram would lie along a straight line.

However, perfect correlation is unlikely to occur in statistical inquiries (though it is common enough in mathematics and scientific theory). Usually we have to deal with weaker relationships – where the plotted points on a dot-diagram are scattered *away* from a straight line. That is to say, the values on one variable keep roughly in step (positively or negatively) with those on the other – but not exactly in step. In general, then, the closer the plotted points lie to a straight line, the stronger the relationship – the higher the degree of correlation. (And the higher the degree of correlation, the more confidently we can predict values of one variable, given values of the other.)

Which of the three scatter diagrams below would seem to show the *highest* degree of correlation?

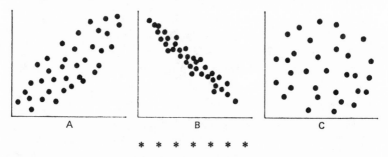

A   B   C

*   *   *   *   *   *   *

The highest degree of correlation is shown in B, where the plotted points lie closest to a *straight* line. In A, the points are more scattered. In C, the points are so scattered that we can't see any sign of a straight line, indicating about zero correlation.

So the amount of scatter in a scatter diagram gives us a rough measure of the strength of a correlation. But for serious statistical work something more precise is required. We need a *numerical* measure – a number, or index, or coefficient that will be at a maximum when the correlation is strong and reduce to a minimum as the correlation weakens. Such an index exists and is called the CORRELATION COEFFICIENT. It is usually represented by the letter r.

Like an average or a measure of dispersion, the correlation coefficient is a *statistic*. It helps describe a sample – in this case a sample of paired values from two different variables. The correlation coefficient measures the closeness with which the pairs of values fit a straight line.

In fact, there are several different kinds of correlation coefficient (as there are, you'll remember, different kinds of averages and measures of dispersion). The one most commonly used is called (because of its mathematical origins, which we won't delve into) the PRODUCT-MOMENT CORRELATION COEFFICIENT. The appropriate formula takes into account the amount by which each value differs from the mean of its own distribution, the standard deviation of the two distributions, and the number of pairs of values.

There is also a *non-parametric* formula which produces a RANK CORRELATION COEFFICIENT. It can be used when measured values do not exist (or are unreliable), especially when sample-members have been merely ranked in order on two category-variables. The closer the relationship between rankings, the bigger will be the correlation coefficient.

Both formulae are so arranged that the correlation coefficient (r) cannot lie outside the range between $+1$ and -1. Those two values of r represent, respectively, perfect positive and perfect negative correlation. When $r = 0$, there is no correlation at all. The closer the correlation coefficient gets to $+1$ or -1, the stronger the correlation; the closer it gets to 0, the weaker it is:

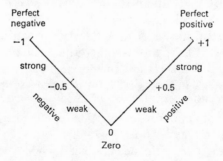

Thus $r = +0.73$ would indicate a stronger relationship than $r = +0.45$; and $r = -0.84$ would be stronger than either of them. The coefficients for the three correlations shown in the scatter diagrams on page 160 are, in (A) $r = -0.67$; in (B) $r =$ approx. 0; and in (C) $r = +0.95$

Look at the three scatter diagrams opposite. The *sizes* of the coefficients for these correlations are 0.98, 0.58 and 0.47. But what *sign* ($+$ or $-$) goes with each coefficient, and which coefficient do you think belongs to each diagram?

*   *   *   *   *   *   *

The correlation coefficients for the relationships shown opposite are: in (A) $r = -0.58$; in (B) $r = +0.47$; and in (C) $r = -0.98$. It may have taken you a while to decide which of (A) or (B) was the more scattered. In fact, it is one of the limitations of scatter diagrams that with correlations of less than about $r = 0.5$, it's quite difficult to judge the difference in scatter.

The significance of a correlation coefficient

So the correlation coefficient describes the strength and direction of the relationship between pairs of values from two different variables. But, of course, it is based upon a *sample*. For example, we may find a correlation of $r = +0.80$ between lateness and travelling-time to college among a sample of 49 students. By now, you'll be sophisticated enough in the ways of statistics to realize immediately that if we took another sample of 49 we'd be most unlikely to get a correlation coefficient of exactly $+0.80$. We might easily get $+0.75$ or $+0.89$, for instance – simply as a result of sampling variation.

So we cannot jump from description to inference and say that, in the population as a whole, $r = +0.80$ also. How far we can trust a sample-r as an estimate of population-r will depend on two factors. The first is the *size* of the coefficient. The bigger it is, the less likely it is to have arisen by chance. We'd be unlikely to find $r = -0.95$, for instance, unless the two variables really are

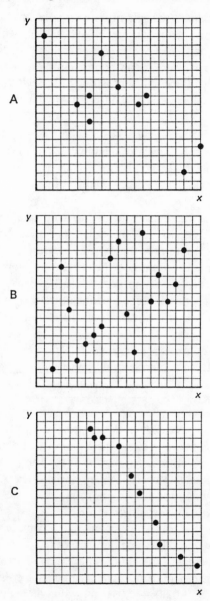

negatively related (though not necessarily so strongly). If, on the other hand, we found $r = +0.10$, say, we'd not be too surprised to get $r = 0$ or even (reversing the direction of the relationship) $r = -0.10$ on another sample.

So the size of the coefficient is one of the two factors on which its reliability depends. Can you imagine what the other will be?

*   *   *   *   *   *   *

The other factor is the size of the sample. The more pairs of values we have in the sample, the more likely we are to see a similar correlation coefficient in other samples (and in the population as a whole).

We can use these two factors to calculate the STANDARD ERROR OF THE CORRELATION COEFFICIENT. (In this case, the standard error represents the standard deviation of a hypothetical distribution of correlation coefficients among samples of a given size drawn from the same population.) We can estimate it by squaring the correlation coefficient, subtracting it from one, and then dividing it by the square root of the number of pairs in the sample.

So, the standard error of the correlation coefficient of $+0.80$ which came from a sample of 49 pairs is:

$$SEr = \frac{1 - (+0.80)^2}{\sqrt{49}} = \frac{1 - 0.64}{7} = \frac{0.36}{7} = 0.05$$

As you can see, *the bigger the correlation coefficient*, the larger would be the number subtracted from 1, and the smaller the number into which the square root of the sample size is divided; and so the *smaller the size of the standard error*. Similarly, *the larger the sample*, the larger the square root we divide by, and hence *the smaller the standard error*.

What does this mean? It means that if we continued to draw samples of 49 from the same population to study the relationship between lateness and travelling-time, we could expect to find that about 68% of them had correlation coefficients in the range $r \pm 1$ SEr, (i.e. $+0.80 \pm 0.05$); the value of r would lie between $+0.75$ and $+0.85$. Furthermore, about 95% of all coefficients would lie between $r \pm 2$ SEr, (i.e. $+0.80 \pm 0.10$). And about 99% would lie within $2\frac{1}{2}$ SEr either side of $+0.80$, (i.e. $+0.80 \pm 0.125$). These latter two ranges are the 95% and 99% confidence intervals for the correlation coefficient we found. To sum up, we can be:

68% sure that population-r lies between $+0.75$ and $+0.85$
95% sure „   „   „   „ $+0.70$ and $+0.90$
99% sure „   „   „   „ $+0.675$ and $+0.925$

If we wanted to be practically certain (99.7%) of including the true correlation coefficient of the population, we would say it lies in the range of $r = +0.80 \pm$ (*how many*?) SEr.

* * * * * * *

The true correlation coefficient of the population (like the true mean of a population) is 99.7% certain to lie within 3 standard errors of the sample statistic; in this case, within the range $+0.80 \pm 3(0.05)$. So the true correlation coefficient for the population from which our sample-r of $+0.80$ is drawn is most unlikely to be smaller than $+0.65$ or larger than $+0.95$.

With a correlation coefficient as large as $+0.80$ and from a sample of 49 pairs, we need have little doubt that the relationship is genuine. It could scarcely have arisen by chance from a population with a true r of 0 (or thereabouts). If our sample-r had been small, however, we could not have been so certain. Suppose, for instance, that we'd found $r = +0.10$ in this sample. This would have given a standard error of

$$\frac{1 - (+0.1)^2}{\sqrt{49}} = \frac{1 - 0.1}{7} = \frac{0.9}{7} = 0.13$$

Clearly, this sample correlation coefficient could easily have arisen in a population with $r = 0$. The standard error is actually larger than the coefficient. Even the range of $r \pm 1$ SEr, giving possible values of r between -0.03 and $+0.23$ (taking in 68% of all correlation coefficients from such samples), would be sufficient to reverse the relationship, by including negative correlations as well as positive ones! We could have no faith here that we'd found a reliable relationship. Though it clearly does exist in the sample, it perhaps does not in the population.

If we want to state the certainty with which we claim to have found a real relationship, we can apply a significance test. (There is, in fact, a t-test in which a t-value is calculated from the size of the coefficient and the size of the sample, and then assessed by reference to the t-distribution mentioned in the previous chapter.) What do you think the null hypothesis would be in testing the significance of a correlation?

* * * * * * *

The null hypothesis is that there is no relationship in the population; thus we assume that population-$r = 0$.

If the population-r were really 0, it would be normally distributed with a standard error of:

$$\frac{1 - (0)^2}{\sqrt{\text{sample-size}}} = \frac{1}{\sqrt{\text{sample-size}}}.$$

So, in about 95% of samples from such a population, sample-r would be within 2 standard errors of 0; and in about 99% it would be within $2\frac{1}{2}$ standard errors. Hence, as a rough-and-ready test (slightly more inclined to Type II errors than the t-test), we can say that a sample correlation coefficient is significant ...

at the 5% level if its size (+ or −) exceeds $2 \times \dfrac{1}{\sqrt{\text{sample-size}}}$

and

at the 1% level if its size (+ or −) exceeds $2\frac{1}{2} \times \dfrac{1}{\sqrt{\text{sample-size}}}$

Thus, in a sample of 49 pairs of values, to be significant at the 5% level, the coefficient would have to be greater than

$$\pm \frac{2}{\sqrt{49}} = \frac{2}{7} = 0.29 \, (+ \text{ or } -)$$

To be significant at the 1% level it would have to exceed

$$\pm \frac{2.5}{\sqrt{49}} = \frac{2.5}{7} = 0.36 \, (+ \text{ or } -)$$

Beyond a doubt, our correlation coefficient of +0.80 was significant not only at the 5% but also at the 1% level. (The probability was clearly far less than one in a hundred that such a large coefficient would have arisen from a population where really $r = 0$.) Equally obviously, the coefficient of 0.10 would not have been significant at either level.

Suppose you calculated the height/weight correlation in a sample of 25 students. What size of correlation coefficient would you need to convince you that the relationship was significant at the 5% level?

* * * * * * *

The coefficient would be significant if it was greater than

$$\pm \frac{2}{\sqrt{25}} = \frac{2}{5} = \pm 0.40$$

To be significant at the 1% level it would have to exceed

$$\pm \frac{2.5}{\sqrt{25}} = \frac{2.5}{5} = \pm 0.50$$

Interpreting a correlation coefficient

As we've seen before, a statistic can be 'significant' without, in the everyday sense, being either strong or important. This is equally true of the correlation coefficient. Quite a small value of r can be significant if it has come from a large enough sample. For example, in a sample of 1,000, a correlation of ± 0.08 would be significant at the 1% level. This leads many people to ask what counts as a weak correlation or a strong one – or even to ask when a correlation can be considered 'satisfactory.'

If you remember our discussion on the usefulness of attaching verbal labels to levels of significance, you won't expect me to be very sympathetic to verbal descriptions of correlations. However, you'll find other writers using them – though not necessarily in the same way as one another. But, just as a guide, here is the kind of label they might attach to particular ranges of the correlation coefficient (whether positive or negative, of course):

0.0 to 0.2   very weak, negligible
0.2 to 0.4   weak, low
0.4 to 0.7   moderate
0.7 to 0.9   strong, high, marked
0.9 to 1.0   very strong, very high

What about the other question, as to what counts as a 'satisfactory' level of correlation? This is as daft as asking what is a satisfactory height. It all depends on the context. A height satisfactory for an aspiring jockey or female ballet dancer might not be satisfactory for an aspiring policeman or tennis champion. Similarly, in some circumstances a small correlation may be highly satisfactory, whereas in others a high one may not be high enough. But perhaps in the majority of cases, the question of 'satisfactoriness' is totally irrelevant.

Consider these cases:

(i) A sociologist establishes a correlation of +0.65 between the ages of men and women getting married in a certain town.

(ii) A mathematics test we devised has been criticized on the grounds that it demands that students read verbal descriptions of problems and so gives an advantage to students with higher verbal ability. To check the validity of this criticism, we try out the math test and also a test of verbal ability on a sample of students and calculate the coefficient of the correlation between the two sets of scores. It is near zero.

(iii) A new method of measuring soil acidity has been proposed. We test a sample of different soils, using both methods, on two different occasions. The correlation between the two measurements using the old method is +0.97. The correlation between the two sets of measurements with the new method is +0.90.

In the above cases, would you think any of the correlations were considered satisfactory or unsatisfactory? If so, which?

*   *   *   *   *   *   *

In (i) it seemed irrelevant to regard the obtained correlation (or any that might have been obtained) as satisfactory or unsatisfactory. It was just a fact. In (ii) the correlation (or lack of it) was considered satisfactory because it indicated that the math test was not being 'contaminated' by testing verbal ability also. In (iii) the correlation obtained by test-retest using the new method, although high, was not high enough to be satisfactory compared with that obtained from the old method.

So the correlation between the values of two variables is a *fact*. Whether we regard it as strong or weak, satisfactory or otherwise, is a matter of *interpretation*. Another matter of interpretation arises when people ask whether one variable somehow causes or determines the other. Now correlation does *not* imply causation. If variables X and Y are correlated, this may be because X causes Y, or because Y causes X, or because some other variable is affecting both X and Y, or for a mixture of these reasons; or the whole relationship may be a coincidence.

For instance, suppose someone points out that there is a strong correlation between adults' intelligence-test scores and the amount of time they have spent being educated. Does this mean that IQ is 'caused by' education? (In other words, can we say that the longer a person spends in education, the larger he can expect his IQ to become?) Possibly, but the relationship could be working the other way round. The higher a person's IQ, the longer he will have chosen (or been allowed) to stay in education. If so, IQ will have 'caused' education. Or it could be that both these variables (IQ of adults and years spent studying) are caused by an outside variable – for example, the intelligence of the parents. That is, the more intelligent the parents, the more intelligent their children are likely to be and the more likely the parents are to encourage them to prolong their education. If such were the case, X and Y would be correlated, not because one caused the other, but because of an outside variable, Z.

Another example: it has been pointed out that there is a high positive correlation between the size of a child's hands and the quality of his handwriting. How would you explain it? Would you say that:

(a) bigger hands can hold a pen more steadily? or
(b) practice in writing makes hands bigger? or
(c) there is a third variable and, if so, what might it be?

* * * * * * *

Of course, there is a third variable (c) affecting both the size of a child's hands and the quality of his writing – his age.

But very often this game of 'hunt the missing variable' will be quite fruitless. It may be that the relationship is a coincidence. For instance, there was a high positive correlation between the proportion of marriages taking place in church and the death-rate during the years 1866 to 1911. That is, in years when people were most likely to get married in church (rather than in a civil ceremony) the death-rate was higher than in years with a smaller proportion of church-weddings. One could rack one's brain for explanations. Did getting married in church cause people to die faster than they would otherwise have done? Or was it the increase in death-rate that made people want a church-wedding more? Or (to be subtle) was it that people who choose to get married in church are somehow weaker and therefore more likely to die than most?

All those explanations are nonsense, of course. The variables are quite unconnected. All they had in common was that both were decreasing over time. During the years 1866 to 1911, improved medical science and public hygiene lowered the death-rate, while the influence of religion weakened and fewer people thought it necessary to get married in church.

The fact is, any two sets of numbers that have an overall trend upwards or downwards are likely to show *some* correlation if you consider them in pairs. There may nevertheless be no sensible, logical connection between the two variables. Correlation (a mathematical relationship) can never prove a causal connection. What it does do is give support to an explanation you can justify on *logical* grounds. And clearly this explanation will carry more weight if you've used it to anticipate the correlation before collecting the data. Hence, correlation is often used in the first stages of testing a theory. By checking whether two or more variables are closely related to one another, the researcher can pin down the factors that explain and control the variable he is interested in.

One of the first theories to be tested in this way was that human beings inherit physical characteristics from their parents. Take, for instance, the theory that the heights of parents somehow determine the heights of children. What kind of correlation would you look for to lend support to this theory? (Negative or positive? Between what and what?)

* * * * * * *

You would look for a high positive correlation between the heights of children (when fully grown) and their parents. And you would find it. But, of course, the correlation would not be perfect, because there is variation within families as well as between them. Besides, would we expect a girl's height to be correlated with that of her mother, and a boy's with that of his father; or any child's to be correlated with some kind of average of both parents' heights, or what? This leads us into a consideration of just what is implied by the *size* of a correlation coefficient.

Suppose we have compared the heights of a sample of sons and their fathers and found the correlation to be $+0.8$. It seems reasonable to say that the sons' heights can, to *some* extent, be 'explained by' the fathers' heights. But where does the 0.8 come in? Does this mean that 80% of a son's height was caused by his father's height? Or that 80 out of every hundred sons owe their heights to their fathers? Or what? What does the size of the correlation coefficient tell us?

For reasons which I won't go into, it is accepted that the *square* of the correlation coefficient tells us how much of the variation in one variable can be explained by variations in the other. Thus, we see that $0.8^2 = 0.8 \times 0.8 = 0.64$ or 64% of the variation in our sample of sons' heights can be explained by variations in their fathers' heights. Thus, 36% of the variation is to be explained by *other* factors.

Similarly, many studies have shown correlations between students' SAT scores and their cumulative averages of less than $+0.4$. If these SAT scores are taken to represent 'aptitude' (as they are by most universities and colleges), it appears that less than *what* percentage of grade average is determined by aptitude, while *what* percent is to be explained by other factors?

$$* \quad * \quad * \quad * \quad * \quad * \quad *$$

It appears that less than 16% of the variation in grade average is determined by aptitude, while 84% is to be explained by other factors (personality, diligence, motivation, luck, etc.).

Let's return to the relationship between the heights of children and parents. We found that the heights of fathers accounted for only 64% of the variation in heights of sons. Clearly, other factors are at work also.

What other factor would you think is the chief one that might help account for the variability in the heights of sons?

$$* \quad * \quad * \quad * \quad * \quad * \quad *$$

Surely, the heights of mothers is the chief other factor. (Diet and exercise could also have some effect.)

In fact, we thought of this before carrying out the investigation. We checked the correlation between the sons' heights and the mothers' heights also. It was +0.7. This means that about *what* percent of the variation in sons' heights can be explained by variations in the heights of mothers?

$$* \quad * \quad * \quad * \quad * \quad * \quad *$$

About 49% of the variation in sons' heights can be explained by variations in mothers' heights.

So the sons' heights were more closely related to the fathers' heights than to the mothers'. If you had to choose, it would be safer to predict a particular son's height from that of his father than from that of his mother.

'But wait a minute,' I hear you say. 'If 64% of the variation in sons' heights is due to the fathers and 49% to the mothers, this adds up to 64 + 49 = 113%, which doesn't make sense.'

A good point, and here is the explanation. Some of the variation in sons' heights is caused by variations in fathers' and mothers' heights *acting together*. There is an overlap or interaction effect. The fact that the 'explained' part of the variation in sons' heights comes to more than 100% is due to the fact that fathers' and mothers' heights are correlated also with *each other*. This is reasonable enough. When taking marriage partners, tall people tend to prefer other tall people, and short people go for short people.

As another example, consider an American study in which the social class of families was correlated with the amount of education of the father; and also with the 'grade' of area in which the family lived. The coefficients of correlation were +0.78 and +0.69 respectively. Were the variables 'grade of living area' and 'amount of education' correlated with each other?

* * * * * * *

Of the variation in social class, $0.78 \times 0.78 = 61\%$ is explained by father's education and $0.69 \times 0.69 = 48\%$ is explained by (*not* caused by) the living area. Since $61 + 48$ is more than 100%, these two latter variables must be correlated with each other. This is surely logical, since better-educated people generally have more money and can live in more 'desirable' areas than poor people.

This technique of explaining variation in a variable in terms of variations in other variables is another aspect of the analysis of variance. It would lead into several techniques (partial correlation, multiple correlation, factor analysis, cluster analysis) which can be of great value to researchers faced with several causes acting at once. However, we can't go any further in this book.

One thing you may have got from this discussion of 'explained' and 'unexplained' variation is some idea of how to weigh up the size of a correlation coefficient. For instance, it's obvious that a correlation of, say, -0.6 between two variables is stronger than if the correlation were only -0.3. But is the relationship *twice* as strong? Do you think that

(a) yes, it is exactly twice as strong? or
(b) no, it is *more* than twice as strong? or
(c) no, it is *less* than twice as strong?

* * * * * * *

No, a correlation of -0.6 is more than twice as strong as one of -0.3. It would mean that $0.6 \times 0.6 = 36\%$ of the variation in one variable could be explained by variation in the other variable, rather than a mere $0.3 \times 0.3 = 9\%$.

In other words, a doubling of the correlation coefficient means a quadrupling of the amount of agreement between the two variables concerned. This in turn means that we are disproportionately safer in making estimates and predictions when r is larger. (Here, for instance, in making estimates when $r = -0.6$ we would be four times more confident (not two times) than if $r = -0.3$).

Actually, trying to estimate or predict from one variable to the other is not a very rewarding business when the correlation coefficient is much below ± 0.80. However, smaller correlations are more helpful than none at all. But this takes us into the field of *regression*.

Prediction and regression

It often happens that we have two sets of related values (for example, the course-work marks of a group of students and their final exam marks), and we want to estimate or predict the value on one variable that would correspond with a given value on the other. For example, we know Sue Smith's course-work mark, but for some reason she was unable to take the final exam. Can we estimate what her mark would have been? Well, if there were no correlation between the two sets of marks, we could play safe by giving her the *mean* of the exam marks scored by her fellow-students. But, if there is any sort of correlation, it will help us to make a better-informed estimate. How much better depends on the strength of the correlation.

Let's say that Sue's course-work mark is 60. Now, suppose we make a scatter diagram for the course-work and exam marks of the other students. We can then see what kind of exam scores were made by other students who scored 60 in course-work. The range of their exam marks gives us an estimate for Sue's.

However, as I indicated above, the precision of this estimate depends on the degree of correlation. For instance, either of the two scatter diagrams below might describe the relationship. In which case (A or B) would it be possible to give the more precise estimate of Sue's exam score?

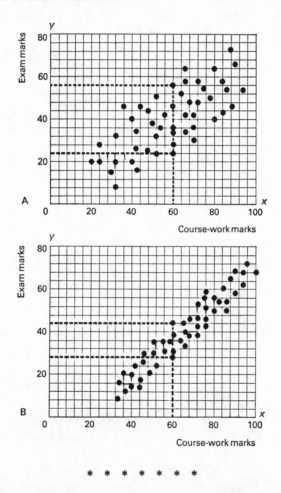

* * * * * * *

The more precise estimate can be made in relationship (B). Reading up the vertical line from a mark of 60 on the course-work scale, we see the students who scored that mark and then read across to see their exam marks. In (A), such students scored between 24 and 56. In (B) they scored between 28 and 44. So, if (A) represented the correlation in Sue's group, we'd have to estimate her exam score as 'between 24 and 56' (a range of 32); but if (B) represented her group, our estimate would be 'between 28 and 44' (thus narrowing the range to 16).

Clearly, the less the vertical spread of the scattered dots, the more precise our estimates; and, of course, the less the spread, the greater the correlation. Perfect (100%) estimates or predictions are possible only when all the dots lie on a straight line (as they did with our example of the radii and circumferences of circles). With perfect correlation, we can say exactly what value of one variable will go with any given value of the other.

Hence, one approach to prediction is to 'reduce' the data to a straight line. We ask: 'What is the "underlying" straight line from which all these points deviate?' We look for the so-called 'LINE OF BEST FIT'. This acts rather like a measure of central tendency, in that it attempts to average out the variations.

Here, for example, is where I'd put the 'line of best fit' on one of the two scatter diagrams we just looked at. I could then use this line to make a precise estimate of the exam mark corresponding to any given course-work mark. If we took this line to represent

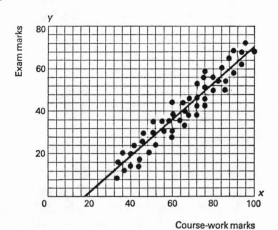

Course-work marks

the relationship, what exam mark would correspond to Sue's course-work mark of 60?

* * * * * *

The corresponding exam score would be 36.

The problem is, of course: where does the line of best fit belong? We can sketch it in 'by eye,' as I did above. In doing so – trying to find a sensible straight path through the dots – we must try to judge whether, on average, the dots would deviate by as much on one side of a possible line as they would on the other. No easy task! And you can be sure that we'd all draw rather *different* lines of best fit, with the result, of course, that we'd all make slightly different estimates of Sue's missing mark. And, again, the weaker the correlation, the bigger the scatter, and the more we'd differ in where we drew the line – see how difficult it would be to decide in the other scatter diagram on page 177 – and so the more we'd differ in our estimates.

Not surprisingly, then, techniques exist for *calculating* a position for lines of best fit. (They must, for example, pass through the mean score on both variables.) Such lines are called REGRESSION LINES. The term was introduced by the nineteenth-century British scientist, Francis Galton. In studying the relationship between the heights of sons and their fathers, he found that, while taller-than-average fathers tended to have taller-than-average sons (and smaller fathers, smaller sons). the sons tended to be nearer the average height of all men than were their fathers. This he called a 'regression to mediocrity' – a falling back towards the average. He and his friend Karl Pearson (who introduced the correlation coefficient) developed a variety of techniques for studying such relationships, and these became known as regression techniques.

Like any other straight line on a graph, a regression line can be described by an equation. This is called a 'REGRESSION EQUATION'. For instance, letting x stand for the course-work marks and y for the exam marks, the regression equation for the line in the diagram above would be $y = \frac{7}{8}x - 17$. Values of x

(the exam marks) can then be estimated directly from the equation without looking at the scatter diagram any further.

Try using the equation yourself. Suppose a student made a course-work mark (x) of 80. What would you estimate as his exam mark (y)?

*   *   *   *   *   *   *

Since exam marks (y) and course-work marks (x) are connected by the regression equation $y = \frac{7}{8}x - 17$, a course-work mark (x) of 80 suggests an exam mark of:

$$y = \tfrac{7}{8}(80) - 17$$
$$= 70 - 17$$
$$= 53$$

As you can check from the scatter diagram, this seems to agree with the value we'd read off, using the regression line directly.

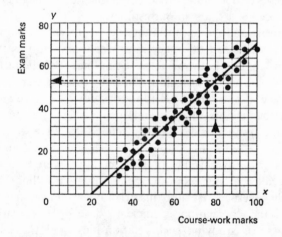

Of course, you may well wonder at the wisdom of making precise-looking estimates on the basis of a regression line, especially when the points are actually scattered quite far away from it. In fact, any such estimate could only be stated with honesty if accompanied by an indication of the possible error – do you remember the idea of a confidence interval?

An alternative approach to prediction is to display the scatter in a table. In our scatter diagrams so far, we have assumed that no two members of the sample had the same pair of values, e.g. that no two students both scored 50 on course-work and 40 in the exam. So each dot has represented a different member of the sample. But, in practice, especially with large samples, it's quite likely that, on several occasions, a particular pair of values would be shared by more than one member of the sample. To represent this diagrammatically, we'd need a three-dimensional chart – with columns instead of dots, varying in height according to the number of times each particular pair of values was observed. Another solution is to use a table like the one below:

Practical and theory marks of 118 students

Theory marks		0	1	2	3	4	5	6	7	8	9	10
	10											
	9									1	1	1
	8						2	4	3	2	1	
	7				1	3	4	3	4	1		
	6			1	3	6	5	4	2			
	5			1	6	6	4	1	1			
	4	1	2	4	6	4	2	1				
	3	1	2	4	2	2	2					
	2	1	3	2	3	2	1					
	1		1	1								
	0											
		0	1	2	3	4	5	6	7	8	9	10

Practical marks

This table shows that, for example, of the ten students who scored 7 in the practical test, three scored 8 on the theory test, four scored 7, two scored 6 and one scored 5. We might then predict, on the basis of this sample, that a student who scored 7 on the practical would have a $\frac{4}{10}$ or 40% chance of scoring 7 in the theory test. We might also point out that, although the mean theory score of students scoring 7 on practical is 6.9, such a student has a 30% chance of scoring 8 on theory.

Again, we might notice that the mean theory score of students scoring 5 on practical is 5.5. However, suppose we think of a wider population of similar students who might have taken the practical test but not the theory. If we can generalize from the sample in this table, what is the probability that such a student who scored 5 on the practical will score (i) less than 5; (ii) more than 5 on the theory test?

* * * * * * *

Of the twenty students who scored 5 on practical, five scored less than 5 on theory and eleven scored more. So the probability of a student with 5 on practical scoring (i) less than that on theory is $\frac{5}{20}$ or 25%, while the probability of his scoring (ii) more is $\frac{11}{20}$ or 55%.

Notice that the table makes clear the evidence on which our prediction is based. While a precise estimate can be offered (e.g. the mean score), the probability of greater or lesser scores is plainly visible.

If we were to draw such a table for a large sample, covering as wide a mark-range (0–100) as we envisaged for the pairs of course-work/exam marks, we'd probably *group* the data. Thus, as you'll see in the example below, one cell might show that 6 students scored *between* 60 and 69 on course-work and *between* 50 and 59 on the exam. Precise information is lost. We no longer show exactly what course-work mark and exam mark each of these students obtained. But the data are easier to handle, and the only precision lost in the estimates we'd make is a fairly spurious kind of precision anyway.

Course-work and final exam marks of 123 students

Final exam marks	0-9	10-19	20-29	30-39	40-49	50-59	60-69	70-79	80-89	90-100
90-100									1	1
80-89								1	3	1
70-79						2	2	4	3	2
60-69				1	1	2	4	5	4	1
50-59				1	2	3	6	3	3	1
40-49				2	4	4	3	2	2	1
30-39			1	3	5	3	4	1	1	
20-29		1	2	2	4	3	1	1		
10-19		1	3	4	2	1				
0-9	1	1	2	1						

Course-work marks

Sue, with her course-work score of 60, would obviously be compared with one of the twenty students in the '60–69' column.

It would seem that there is only a $\dfrac{4 + 2}{20} = \dfrac{6}{20} = 30\%$ chance

of her having made the same score or higher on the final exam. Indeed, there is a $\frac{4}{20}$ or 20% chance that she may have scored as low as 30–39. There is even a $\frac{1}{20}$ or 5% chance that her exam score may be between 20 and 29. What is her most likely exam score; and what is its probability?

$$* \quad * \quad * \quad * \quad * \quad * \quad *$$

Sue's most likely exam score is 50–59 marks. Of the twenty students comparable with her on course-work marks, more (6) fell into that exam-mark range than into any other – the probability of such a score is $\frac{6}{20}$ or 30%. (Sue also benefits from the way we happen to have grouped the marks. If she'd scored 59 rather than 60, she'd have fallen in the 50–59 class-work group, and her most likely exam score would be only 40–49.)

Don't forget, however, that we have a very limited sample: only twenty students in Sue's range of course-work marks. Yet we wish to generalize to such students in general, and to Sue in particular. Sue, for all we know, may have special qualities that would enable her to score 100 in the exam. It appears improbable, but it remains possible.

This seems a suitable point at which to begin drawing this book to a conclusion. For we are back where we began, distinguishing between samples and populations, and asking how reasonable it is to generalize from one to the other. And, in any case, what are the risks in predicting for just one individual (Sue) from the population? Statistics, as I'm sure you'll have gathered, is far better equipped to make inferences about things 'in general' and 'in the long run' than about particular things on a particular occasion. This, though, is one of its most important lessons. Human beings are bound to generalize, expecting to see the characteristics, differences, trends and associations we have noted in one instance to be repeated in apparently similar instances. Without some such expectation of constancy, we could not survive. But all such expectations must be hedged about with probability. They must be tentative rather than absolute. We must expect them to be confounded much of the time. Unless we can, to some extent at least, expect things to be different from what we expect, then we cannot learn from experience. And that is also a path to stagnation and extinction.

Postscript

In the preceding pages I've been very much aware that we've been skimming the surface of most issues. This is because my aim was to give you a bird's-eye-view of the field of statistical concepts, rather than take you crawling, like a calculating snake, through the undergrowth of statistical computation. If you feel I've raised more questions in your mind than I've answered, I shan't be surprised or apologetic. The library shelves groan with the weight of the books in which you'll find answers to such questions (see the Bibliography, pages 191–5). Even though you'll be well aware that statistics is a subject with ramifications and inter-connections far beyond anything we've discussed here, you should be pretty confident that you've 'got the feel' of its main lines of concern.

If you have worked through the book to this point, you should now have a basic grasp of the concepts and terminology of statistics. It should be of great help to you in: (1) reading a research report (or a newspaper) with some expectation that you'll see the point of any statistics it may contain; (2) describing your research interests to a professional statistician in terms that would enable him to offer you technical advice; and (3) turning to books or courses on statistical calculation with a view to learning the techniques yourself.

Unless you have to pass an exam in statistics, you may well need little more than you have already learned. Although students in many subjects must take courses in statistics, much of what they are taught often seems of doubtful relevance to their other activities. Certainly research has shown (if it can be relied upon!) that social scientists tend to retain only a partial knowledge of the statistics they were taught. But perhaps they remember all that is useful to them. For when social scientists do make liberal use of statistics in their research reports, one often gets the impression they are being used to make the work look more 'scientific' (and therefore believable) rather than to clarify its meaning.

To conclude, I'll briefly review the 'bird's-eye-view' and add a final note of caution.

Review

Statistics is a means of coming to conclusions in the face of uncertainty. It enables us to recognize and evaluate the errors involved in quantifying our experience, especially when generalizing from what is known of some small group (a sample) to some wider group (the population).

Statistical analysis begins in a description of the sample. We may find diagrams a revealing way of describing a sample and comparing it with other distributions. But we are particularly interested in getting a measure of the sample's central tendency and, where appropriate, a measure of its dispersion. The two most important such measures (with quantity-variables) are the arithmetic mean and the standard deviation. These are of particular value in defining the normal distribution: a symmetrical, bell-shaped curve.

Once we have the mean and the standard deviation, we can compare values from two different distributions (in z-units) and we can estimate the percentage of observations in a distribution that would fall above and below various values of the variable. We can also infer parameter values (in the population) based upon the statistics (from the sample), using the concept of standard error to decide the confidence interval within which we believe the true population value (e.g. a mean or a proportion) to lie. We would commonly quote a range in which we are 95% certain the true value lies, or a wider one about which we can be 99% confident.

Following the same principles, we can compare two (or more) samples, and ask whether they are sufficiently similar to have come from the same population. Or is the difference between them big enough to signify a real difference in populations – one that would be repeated in subsequent pairs of samples chosen in the same way? We make a null hypothesis that the samples come from the same population and that the difference has arisen purely by chance. Tests enable us to determine the plausibility of this hypothesis. If the probability of getting two samples so different from one population is less than 5%, we may reject the hypothesis. If we want to be more careful still, we may prefer not to reject the null hypothesis (i.e. not recognize the difference as a real one) unless it is so big that its probability in two samples from the same population is less than 1%. Such differences are said to be significant (even though they may not be important in any practical sense).

There are also tests to establish whether there appears to be a significant difference anywhere among a group of more than two samples. This involves the analysis of variance: comparing the variance between groups with the variance within groups. When dealing with categories rather than with quantity-variables, and asking whether there is a significant difference between two samples in proportions rather than means, we use a non-parametric technique called the chi-square test. This compares the frequency with which we'd expect certain observations to occur, if chance only were operating, with the frequency that actually occurred. (Such non-parametric techniques are essential when dealing with category-variables and may in other cases be advisable when we can't be sure that the parent population is normally distributed.)

Finally, we are often interested in the relationship between related pairs of values from two different variables (for example, people's heights and weights) Correlation is a measure of such relationship and the correlation coefficient indicates its strength– on a scale from −1 and +1 (equally strong) down to zero. Scatter diagrams are a useful way of displaying correlation, but may need to be replaced by tables when the same pair of values is recorded for several members of the sample. With regression techniques, we can use the relationship observed in the sample to predict, for the population, values of one variable that would correspond with given values of the other variable. The likely accuracy of such predictions increases with the strength of the correlation. Even when the correlation is very strong and predictions are firm, we cannot use that fact to *prove* that one variable causes the other, even if we can explain a causal connection. Variable X may cause variable Y, or vice versa, or both may be determined by another variable, Z, or the mathematical relationship may be a coincidence. As usual in statistics, however, the data would *lend support* (or deny it) to a reasoned argument along one of these lines, but absolute proof is never forthcoming.

Caution

Finally, a few words of caution. It has been pointed out, over and over again, that: 'There are lies, damned lies, and statistics!' or 'Figures don't lie but liars use figures!' or, less hysterically (but despite what I said at the end of the last paragraph) 'You can prove anything with statistics.' *How to Lie with Statistics* by Darrell Huff and Irving Geis (Norton, 1954) is the classic book on this topic; the warnings it conveys are both deadly serious and amusingly presented.

In general, it is well to remember that a person who uses statistics may be someone with an axe to grind. He may be propping up a weak case with rather dubious figures which he hopes will impress or intimidate any potential critics. For example, a British politician recently claimed that '50% more teachers considered that educational standards had fallen rather than risen over the previous five years.' This pronouncement is worrying but rather obscure: 50% more . . . than what? In fact, it was based on a survey in which 36% of the teachers believed that standards of pupil achievement had fallen, 24% believed they had risen, 32% believed they had remained the same, and 8% didn't know. Clearly, the politician ignored the 'stayed the same' and the 'don't know' groups (a cool 40% of the sample) to arrive at his '50% more' figures. Had he wished to propagate a rosier view of British educational standards, he might, with equal validity, have pointed out that no less than 64% of teachers do not believe that standards have fallen. On the other hand, he might have painted an even blacker picture by saying that 76% of teachers do not believe standards have risen!

But please don't think that untrustworthy statistics are found only in the rantings of politicians and the cajolements of advertisers. Just this week, as I put the finishing touches to this chapter, the leading French opinion poll stands accused of suppressing the fact that 77% of the French people polled thought that immigrants should be sent home; and publishing instead the false figure of 57% – a figure that would be more acceptable to their clients (the French government).

Even scientific researchers are subject to human failings. They are most unlikely to attempt out-and-out deception – though the life-work of a famous, recently deceased British psychologist is currently alleged by some critics to be based on fraudulent experimental and statistical data. But even they may get carried away by the desire to 'prove' a pet theory and, in the process, overlook the inadequacy or bias in a particular sample, or choose inappropriate tests or insufficiently stringent levels of significance. Studies of journal articles (in the psychology field, for example) have indicated many such lapses. Indeed, a 'classic' 1971 study by a Finnish medical researcher (which suggested that one's chance of heart disease would be reduced by eating vegetable margarine rather than butter) is now being attacked for just such statistical weaknesses.

Of course, you cannot be expected to sniff out the transgressors. You simply do not have sufficient technical expertise. Nor do I want to encourage you to be unduly suspicious. Your problem will chiefly lie in understanding and interpreting people's statistics, not in trying to catch them. At the same time, your interpretation will be colored by what you have learned in this book: for instance the possibility of bias in samples, the distinction between significance and importance, the fact that correlation does not imply causation, and so on. Needless to say, if you should find yourself producing statistics for other people, to back up an argument of your own, I assume you will strive for all the honesty you'd wish for from others.

In short: as a consumer of statistics, act with caution; as a producer, act with integrity.

Bibliography

Here is a small (and not at all random!) sample from the vast population of books in which you might follow up what you have learned so far. I begin with a small 'general' section, listing books that should be helpful to the reader who wants to go a little bit further. I follow this with sections listing books that apply statistical thinking to each of several particular subject-areas. This is not to imply, however, that such books could be of help only to students of the subject under which they are listed. You might well find the exposition of some authors to be so good that you'd be prepared to overlook the fact that their examples were not drawn from your subject. (In any case, my classification has to be somewhat arbitrary, since some of the books could appear in more than one category.) However, what I am assuming really is that, having endured my general introduction, you'll be most concerned to hear from authors who tackle problems in your own subject-area.

General Introductions

Freund, J. E. *Statistics: A First Course*, 3rd ed. Englewood Cliffs, N.J.: Prentice-Hall, 1981.

Haber, A., and Runyon, R. P. *General Statistics*, 3rd ed. Reading, Mass.: Addison-Wesley, 1977.

Huff, D., and Geis, I. *How to Lie with Statistics*. New York: Norton, 1954.

Sanders, D. H., and others. *Statistics: A Fresh Approach*. New York: McGraw-Hill, 1979.

Behavioral Sciences, Psychology and Education

Crocker, A. C. *Statistics for Teachers*. Atlantic Heights, N.J.: Humanities Press, 1974.

Gehring, R. E. *Basic Behavioral Statistics*. Boston: Houghton Mifflin, 1978.

Gellman, E. S. *Statistics for Teachers*. New York: Harper & Row, 1973.

Guilford, J. P., and Fruchter, B. *Fundamental Statistics in Psychology and Education*, 6th ed. New York: McGraw-Hill, 1977.

Hardyck, C. D., and Petrovorich, L. F. *Introduction to Statistics for Behavioral Sciences*, 2d ed. New York: Holt, Rinehart & Winston, 1976.

Lewis, D. G. *Statistical Methods in Education*. New York: International Publications Service, 1967.

Lynch, M. D., and Huntsberger, D. V. *Elements of Statistical Inference for Psychology and Education*. Boston: Houghton Mifflin, 1976.

MacCall, R. B. *Fundamental Statistics for Psychology*, 2d ed. New York: Harcourt Brace Jovanovich, 1975.

Popham, W. J., and Sirotnik, K. A. *Educational Statistics*, 2d ed. New York: Harper & Row, 1973.

Slakter, M. J. *Statistical Inference for Educational Researchers*. Reading, Mass.: Addison-Wesley, 1972.

Siegel, S. *Nonparametric Statistics for the Behavioral Sciences* New York: McGraw-Hill, 1972.

Business and Management

Braverman, J. D., and Stewart, W. C. *Statistics for Business and Economics*. New York: John Wiley, 1973.

Broster, E. J. *Glossary of Applied Management and Financial Statistics*. New York: Crane-Russak, 1974.

Levin, Richard I. *Statistics for Management,* 2d ed. Englewood Cliffs, N.J.: Prentice-Hall, 1981.

Shao, S. *Statistics for Business and Economics*, 3d ed. Columbus, Oh.: Merrill, 1976
Thirkettle, G. L. *(Wheldon's) Business Statistics*. Philadelphia: International Ideas, 1972.

Economics

Beals, R. E. *Statistics for Economists*. Chicago: Rand McNally, 1972.
Davies, B., and Foad, J. N. *Statistics for Economics*. Exeter, N.H.: Heinemann, 1977.
Jolliffe, F. R. *Commonsense Statistics for Economists and Others*. Boston: Routledge & Kegan Paul, 1974.
Thomas, J. J. *Introduction to Statistical Analysis for Economists*. New York: John Wiley, 1973.

Geography

Ebdon, D. *Statistics in Geography*. Totowa, N.J.: Biblio Dist., 1977.
King, L. J. *Statistical Analysis in Geography*. Englewood Cliffs, N.J.: Prentice-Hall, 1969.
Norcliffe, G. B. *Inferential Statistics for Geographers*. New York: Halstad Press, 1977.

History

Dollar, C. M. *Historian's Guide to Statistics*. Huntington, N.Y.: Krieger, 1974.
Floud, R. *Introduction to Quantitative Method for Historians*. New York: Methuen, 1973.

Medical and Biological Sciences

Brown, B. W. *Statistics: A Biomedical Introduction.* New York: John Wiley, 1977.

Colquhoun, D. *Lectures on Biostatistics.* New York: Oxford University Press, 1971.

Hill, A. B. *A Short Textbook of Medical Statistics*, 10th ed. Philadelphia: Lippincott, 1977.

Hills, M. *Statistics for Comparative Studies.* New York: Methuen, 1974.

Mather, K. *Statistical Analysis in Biology.* New York: Methuen, 1972.

Mosiman, J. E. *Elementary Probability for Biological Sciences.* New York: Appleton Century Crofts, 1968.

Physical Sciences and Technology

Bury, K. V. *Statistical Models in Applied Science.* New York: John Wiley, 1975.

Chatfield, C. *Statistics for Technology*, 2d ed. New York: Methuen, 1979.

Eckschlager, K. *Errors, Measurements and Results in Chemical Analysis.* New York: Van Nostrand, 1969.

Hald, A. *Statistical Theory with Engineering Applications.* New York: John Wiley, 1952.

Johnson, N. L. *Statistics and Experimental Design in Engineering*, 2d ed. New York: John Wiley, 1977.

Koch, G. S., and Link, R. F. *Statistical Analysis of Geological Data.* New York: John Wiley, 1970.

Till, R. *Statistical Methods for the Earth Scientist.* New York: Halstad Press, 1978.

Young, H. D. *Statistical Treatment of Experimental Data.* New York: McGraw-Hill, 1962.

Sociology and Social Sciences

Anderson, T. R., and Zelditch, M. *Basic Course in Statistics with Sociological Applications*, 2d ed. New York: Appleton Century Crofts, 1968.

Blalock, H. M. *Social Statistics*, 2d ed. New York: McGraw-Hill, 1972.

Davis, J. A. *Elementary Survey Analysis*. Englewood Cliffs, N.J.: Prentice-Hall, 1971.

Hays, W. L. *Statistics for the Social Sciences*, 2d ed. New York: Holt, Rinehart & Winston, 1973.

Levin, J. *Elementary Statistics in Social Research*, 2d ed. New York: Harper & Row, 1977.

Ott, L., and others. *Statistics: A Tool for the Social Sciences*, 2d ed. N. Sciuate, Mass.: Duxbury Press, 1978.

Moser, C. A., and Kalton, G. *Survey Methods in Social Investigation*, 2d ed. New York: Basic Books, 1972.

Mueller, J. H. *Statistical Reasoning in Sociology*, 3d ed. Boston: Houghton Mifflin, 1977.

Palumbo, D. J. *Statistics in Political and Behavioral Science*, 2d ed. New York: Columbia University Press, 1977.

Tufte, E. *Data Analysis for Politics and Policy*. Englewood Cliffs, N.J.: Prentice-Hall, 1974.

Index